═══SPACE═══
The Next Twenty-Five Years

Revised and Updated

The Wiley Science Editions

SPACE

The Next Twenty-Five Years

Revised and Updated

Thomas R. McDonough

WILEY

WILEY SCIENCE EDITIONS

John Wiley & Sons, Inc.
New York • Chichester • Brisbane • Toronto • Singapore

Publisher: Stephen Kippur
Editor: David Sobel
Managing Editor: Ruth Greif
Editing, Design, and Production: G&H SOHO, Ltd.

Library of Congress Cataloging-in-Publication Data

McDonough, Thomas R.
  Space, the next twenty-five years. Revised and Updated.

  1. Astronautics. 2. Outer space—Exploration.
I. Title. II. Title: Space, the next 25 years.
TL790.M37 1987   500.5   86-32407
ISBN 0-471-85671-1
      0-471-50335-5 (pbk)

Printed in the United States of America

89 90 10 9 8 7 6 5 4 3 2 1

To the seven who gave their lives so that we might reach the stars,
the astronauts of the Space Shuttle Challenger:
Francis R. (Dick) Scobee, Michael John Smith,
Ellison S. Onizuka, Judith Arlene Resnick, Ronald Erwin McNair,
S. Christa McAuliffe, and Gregory Bruce Jarvis

Acknowledgments

I wish to thank my editor, David Sobel, and his assistant, Dawn Reitz, Ruth Greif, managing editor, and my agent, Sharon Jarvis, for helping to get this book off the launch pad. Julie Glass, my developmental editor, did a particularly superb job of helping to assemble the bits and pieces so the whole vehicle worked without a major malfunction.

I wish to thank the following people for useful conversations and other help in preparing this book: Toni Acker (Toni Acker Associates), Charlene Anderson (The Planetary Society), William Bahr (AT&T), Dennis M. Cole (Lockheed), Jim Davidson (Space Services), Dr. Robert L. Forward (Hughes Aircraft Company Research Laboratories), Dr. Louis Friedman (The Planetary Society), Dr. Martin Harwit (Cornell University), Dr. Neal Hulkower (TRW), Jon Lomberg, Sen. Spark M. Matsunaga, Lyndine McAfee (The Planetary Society), Von Marschall (Messerschmidt-Boelkow-Blohm GmbH), Dennis Meredith and the staff of the News Bureau (Caltech), Dr. Bruce C. Murray (Caltech), Mary Dorothy O'Neill, Jean-Paul Paille (ESA), Jonathan V. Post, Dr. Jerry E. Pournelle, Jeane Rae, Dr. Nancy Rallis (Boston College), Prof. Leah Rynasko (William Carey International University), Dr. Harrison H. Schmitt, Dr. S. Fred Singer (George Mason University), Sandy and Bill Taylor (Reif, Taylor and Taylor), Joan Tuzzolino, and Dr. Jerome Wiesner (MIT).

Thanks to the following Jet Propulsion Laboratory people for much help: Dr. Lew Allen, Valerie Bailey, Dr. Edward Belbruno, Frank Bristow, Dr. Pamela Clark, Dr. Samuel Gulkis, Joel K. Harris, Dr. Albert Hibbs, Dr. Kenneth Jones, Ross Jones, Dr. Gail Klein, Dr. Michael J. Klein, Dr. Nicholas Renzetti, Gael Squibb, Joseph W. Stockemer, Sherry Wheelock, James Wilson, Jurrie van der Woude, Dr. Donald K. Yeomans, and particularly, James D. Burke.

And thanks, especially, to the Sophie T. McDonough Clipping Service.

═══ Acknowledgments for the New Edition ═══

The author wishes to give thanks for help in preparing the revised edition to Steven Aftergood (Committee to Bridge the Gap), Dr. Mark Allen (JPL/Caltech), Dr. Edward Belbruno (JPL), Geoff Biddle (Orbital Sciences Corp.), Dr. David Brin, Prof. Rodney A. Brooks (MIT), Congressman George E. Brown, Jr. and his legislative assistant, Steven M. Wolfe, Dr. James D. Burke (JPL), F. Raul Colomb (Instituto Argentino de Radioastronomia), Sandra Dailey (Quinlan Press), James Doyle (JPL), Robert C. Finn (Caltech), Dr. Louis D. Friedman (The Planetary Society), Bruce Gary (JPL), Dr. Michael Hyson (HyTech), Robert Jamieson (JPL), Dr. Kenneth Jones (JPL), Ross M. Jones (JPL), Dr. Jordin T. Kare (Lawrence Livermore National Laboratory), Dr. Lynn Griffiths (NASA), Dr. Paul Feldman (Johns Hopkins), Prof. Paul Horowitz (Harvard), Guillermo A. Lemarchand (CECEN, Argentina), Dr. Jaume Llibre (Autonomous University of Barcelona), Prof. Bruce C. Murray (Caltech), Prof. Leik Myrabo (RPI), Mel Oleson (Boeing Aerospace), Mary Dorothy O'Neill, Bruce Parham (JPL), Jonathan V. Post (Rockwell), Margaret Power (BD Systems), Dr. Keith Raney (Canada Centre for Remote Sensing), Dr. Nicholas Renzetti (JPL), Dr. Guenter Riegler (NASA Headquarters), Ron Sauder (Johns Hopkins), Ann Shoben (RAND Corp.), S. Fred Singer (DOT), Patricia J. Smiley (NRAO), Donna Stevens (The Planetary Society), Donna Sutton (NASA Headquarters), Ann Tavormina (JPL), Larry Thomas (NASA/GSFC), Paul Turner (Rockwell), Althea Washington (NASA), Randall Weatherington, Alan Wood (JPL), Jurrie van der Woude (JPL), Moreno White (SPARTA), and Tom Williams (McDonnell Douglas). And thanks to the public information offices of the European Southern Observatory (Garching bei Muenchen), The European Space Agency (Washington), General Dynamics (Convair Division), the Johnson Space Center, JPL, NASA Headquarters, and the NASA/Ames Research Center.

Special thanks to my hard-working assistant, Sandy Taylor (Reif, Taylor, and Taylor, Inc.), and to the very able Wiley staff: William Anderson, Ruth Greif, Stephen Kippur, Cathy Pawlowski, David Sobel, Nancy Woodruff, and David Sassian (G&H SOHO, Ltd.). And belated thanks to Claire McKean (G&H SOHO, Ltd.) for her excellent work on the first edition.

And, as always, Sophie T. McDonough.

Permissions

Figures

Page 10: U.S. Army.

Pages 21, 23, 47, 115, 131, 140, 146, 151, 156, 159, 174, 202, 218, 226: NASA/JPL.

Pages 32 (a), 42 (b): European Space Agency.

Page 32 (b): American Rocket Company.

Page 32 (c): Space Services, Inc.

Page 33 (d): Pacific American Launch System, Inc.

Page 42 (a): MBB-ERNO.

Pages 40, 83 (c), 86: Boeing.

Pages 46, 71, 77, 94, 216: NASA.

Pages 55, 65: U.S. Department of Defense.

Page 73: Space Industries, Inc.

Page 82 (a): McDonnell-Douglas.

Page 83 (b): Rockwell.

Page 102: Electromagnetic Launch Research, Inc.

Page 108: © Edward Belbruno, JPL, from a color painting. Permission of the artist.

Page 114: Courtesy of The Planetary Society and the Space Research Institute, Moscow.

Page 128: Joel Hagen © 1986.

Page 168: (Francesco Paresce, Christopher Burrows) Space Telescope Science Institute, operated for NASA by AURA.

Page 171 (a): Courtesy NRAO/AUI, observed by F. Yusef-Zadeh, M.R. Morris, D.R. Chance.

Page 171 (b): European Southern Observatory, observed by Michael Rosa.

Page 181: © 1986 Newspaper Enterprise Association, Inc. Reprinted by permission of NEA, Inc.

Page 187 (a): SPARTA.

Page 187 (b): Orbital Sciences Corp.

Page 188: Rodney A. Brooks, MIT.

Page 190: James D. Burke and Robert S. Jamieson, JPL.

Page 194: Artist Ray Rue for RPI.

Page 197 (a): Martin Marietta Corp.

Page 197 (b): General Dynamics.

Page 205: Hughes Aircraft Company Research Laboratories. Permission of Dr. Robert L. Forward.

Page 214: National Astronomy and Ionosphere Center, Cornell University.

Page 224: © 1986 Jon Lomberg, reprinted by permission of the artist and the artist's agents, Scott Meredith Literary Agency, Inc., 845 Third Ave., New York, NY 10022.

Major Excerpts

Aldrin, Edwin, Pasadena *Star-News* 12/8/88.

Asimov, Isaac, *The Impact of Science on Society*, 1985, NASA SP482, U.S. Government Printing Office, Washington, DC.

Bergman, Jules, *The Impact of Science on Society*, 1985, NASA SP482, U.S. Government Printing Office, Washington, DC.

Bethe, H.A., Boutwell, J., and Garwin, R.L., from *Weapons in Space*, ed. by F.A. Long, D. Hafner, and J. Boutwell, 1986, W.W. Norton, New York. Permission of H.A. Bethe.

Bradbury, Ray, *Why Man Explores*, 1976, NASA EP123, U.S. Government Printing Office, Washington, DC.

Burke, James, *The Impact of Science on Society*, 1985, NASA SP482, U.S. Government Printing Office, Washington, DC.

Bush, Vannevar, quoted in *Facts and Fallacies*, by Chris Morgan and David Langford, 1981, St. Martin's Press, NY.

Chalfont, Alun, *Star Wars*, copyright © 1985. By permission of Little, Brown and Company, Boston.

Clauser, F.H., Griggs, D.T., and Ridenour, L., Project RAND Report SM-11827, 1946, reprinted in *Rand 25th Anniversary Volume*, The Rand Corp., Santa Monica, CA. Permission of The Rand Corp.

Collins, Michael, *Liftoff*, 1988, Grove Press, New York.

Compton, W. David, and Benson, Charles D., *Living and Working in Space*, 1983, NASA Special Publication 4208, U.S. Government Printing Office, Washington, DC.

Connors, Mary M., Harrison, Albert A., and Akens, Faren R., *Living Aloft*, 1985, NASA SP483, U.S. Government Printing Office, Washington, DC.

Cousins, Norman, *Why Man Explores*, 1976, NASA EP123, U.S. Government Printing Office, Washington, DC.

Cousteau, Jacques, *Why Man Explores*, 1976, NASA EP123, U.S. Government Printing Office, Washington, DC.

Drexler, K. Eric, *Engines of Creation*, 1986, Anchor Press/Doubleday, New York.

Dyson, Freeman, oral presentation, 1986. Courtesy of The Planetary Society.

Dyson, Freeman, *Infinite in All Directions*, 1988, Harper & Row, New York.

Ehricke, Krafft, from *Lunar Bases and Space Activities of the 21st Century*, 1985, ed. by W.W. Mendell, Lunar and Planetary Institute, Houston. Permission of Lunar and Planetary Institute.

Flohn, H., *Life on a Warmer Earth*, 1981, International Institute for Applied Systems Analysis, Laxenburg, Austria. Permission of IIASA.

Forward, Robert L., "Feasibility of Interstellar Travel: A Review," International Astronautical Congress, 1985, paper IAA-85-489. Permission of Robert L. Forward.

Forward, Robert L., and Davis, Joel, *Mirror Matter*, 1988, John Wiley & Sons, New York.

Friedman, Louis, *Starsailing*, 1988, John Wiley & Sons, New York.

Froehlich, Walter, *Space Station*, NASA EP213, U.S. Government Printing Office, Washington, DC.

Goldsmith, Donald, *Nemesis*, 1985, Walker, New York. Permission of Donald Goldsmith.

Graham, Daniel O., *We Must Defend America*, 1983, Regnery Gateway, Inc., 950 North Shore Drive, Lake Bluff, IL 60044.

Harwit, Martin, *Cosmic Discovery*, 1981, Basic Books, Inc., NY. Permission of Martin Harwit.

Jones, Eric M., and Finney, Ben R., from *Lunar Bases and Space Activities of the 21st Century*, 1985, ed. by W.W. Mendell, Lunar and Planetary Institute, Houston. Permission of Eric M. Jones and the Lunar and Planetary Institute.

Joyner, Christopher C., and Schmitt, Harrison, from *Lunar Bases and Space Activities of the 21st Century*, 1985, ed. by W.W. Mendell, Lunar and Planetary Institute, Houston. Permission of Harrison Schmitt and the Lunar and Planetary Institute.

Michener, James, *Why Man Explores*, 1976, NASA EP123, U.S. Government Printing Office, Washington, DC.

Contents

SPACE
The Next Twenty-Five Years

Revised and Updated

INTRODUCTION
Space, the Final Frontier

Two years have passed since the first edition of this book appeared, two years during which, outwardly at least, nothing much seemed to be happening in the American space program. Yet remarkable developments were taking place.

The Space Shuttle, grounded in the wake of the *Challenger* disaster, had caused our launches to grind to a halt. The Shuttle went through some wrenching changes and emerged far better for it. This also forced the redevelopment of expendable rockets, so that we will never again be so completely dependent on one class of launch vehicle.

While the Shuttle was on hold, astronaut Sally Ride led a group for NASA that proposed goals for the American space program, independently making recommendations similar to those in the first edition of this book. NASA then formed the Office of Exploration to study these goals, and by the end of 1988 they had proposed four major, far-reaching recommendations:

- A research station on the Moon for scientists, which would lead to
- A permanent, inhabited lunar base, one mission of which would be to help prepare
- A human expedition to the Martian moon Phobos. John Aaron, head of the Office, suggested we could land there as early as 2003. This would lead to
- A human expedition to the surface of Mars. Aaron thinks this might occur by 2007. Soviet participation was suggested for both Martian missions.

In the meantime, the Soviets haven't been standing still. They've launched Energia, their biggest successful rocket ever. They've flown their own space shuttle, have sent two probes to Mars (one designed to hop around its moon Phobos like a frog), and are steadily working on an ambitious program of Martian invasion culminating, they now openly claim, in the flight of cosmonauts to the appropriately red planet. In 1988, they set yet another world record when two cosmonauts spent a year in space.

Glasnost has infiltrated the space program, too. Never before has there been such openness about Soviet space plans or so much cooperation with the West. American scientists participated in the Soviet Phobos mission at Mars, and Soviet scientists are working with the American *Mars Observer*, to be launched in 1992. And much discussion is taking place about possible Soviet-American robotic and even human missions to Mars. (The extent of *glasnost* may be measured by the fact that Soviet cosmonauts requested that the Pink Floyd song "Delicate Sound of Thunder" be played at a recent launch.)

With the liftoff of the Shuttle *Discovery* in September of 1988, the NASA space program was once again rolling and promising the launch of some of the most sophisticated spacecraft ever, making the next several years among the most exciting in the history of the exploration of the universe.

Now that the NASA Shuttle is flying again, the *Galileo* probe to Jupiter will be launched, taking a circuitous route by way of Earth, Venus, and two asteroids. It will also spy on our own Moon, looking down into its pole for the first time, where ice may be hidden—a valuable resource for space settlements. It will for the first time study the Earth from space the way a visiting alien would, taking pictures and making measurements as it approaches.

The *Magellan* spacecraft will fly to Venus, yielding photographic-quality radar images of that shrouded planet.

The *Voyager* probe will take the first ever closeup pictures of Neptune and its mysterious moon Triton.

The *Ulysses* spacecraft, a joint project of the European Space Agency and NASA, will be launched, eventually flying over the poles of our Sun—exploring the part of our star we have never seen clearly and giving clues to its behavior that will help us predict what the climate of the Earth will be in the future.

The Hubble Space Telescope, a project of NASA and the European Space Agency, will be launched soon, giving us an instrument half the diameter of the Mt. Palomar mirror, the first large optical astronomical telescope ever put in orbit. Peering into the farthest reaches of the universe ever seen, we will also be looking back to the beginning of time. To me, this is one of the most exciting developments in the history of science, and if that history is any guide, it will reveal sights no human has ever dreamed of. It could revolutionize our conception of the universe as much as Galileo's telescope did when it was first turned on the heavens.

Historically, whenever we have opened up a new window on the universe, as we did with optical telescopes, infrared detectors, radiotelescopes, and neutrino detectors, we have found startling phenomena that changed our understanding of nature: new planets, distant galaxies, quasars, pulsars, giant molecular clouds, black holes. In the last few years, it has become clear that 90 percent of the universe is *missing*—that all this mass takes the form of some kind of dark matter that we have never been able to detect directly. Ninety percent! The dark matter can only be detected indirectly, by its gravitational effect on other bodies. All the astronomy we've done, from the first cave man or woman who looked up into the night and wondered about the stars, down to the present day, has been contemplating a mere 10 percent of the universe. What is the rest? Nobody knows!

Hubble is just one window in NASA's efforts to learn more about our universe. Their Astrophysics Program has a wonderful plan to explore the universe using almost the entire electromagnetic spectrum, from radio wavelengths to gamma rays, thus to see it through virtually every window allowed by the laws of physics. With aircraft, atmospheric rockets, balloons, and spacecraft, they will blanket the sky. The list of proposed space missions and launch dates includes:

- Hubble Space Telescope (HST, with ESA, 1989)
- Cosmic Background Explorer (COBE, 1989)
- Gamma Ray Observatory (GRO, with West Germany, 1990)
- Roentgen Satellite (ROSAT, X-ray, with West Germany and England, 1990)
- Broadband X-Ray Telescope (BBXRT, 1990)
- Astro 1 (ultraviolet telescope, 1990)
- Extreme Ultraviolet Explorer (EUVE, 1991)
- Orbiting and Retrievable Far and Extreme Ultraviolet Spectrometer (ORFEUS, with West Germany, 1992)
- Diffuse X-Ray Spectrometer (DXS, 1992)

- Astro D (X-ray, with Japan, 1993)
- Infrared Telescope in Space (IRTS, with Japan, 1993)
- X-Ray Timing Explorer (XTE, 1994)
- Advanced X-Ray Astrophysics Facility (AXAF, with Holland, 1995)
- Space Infrared Telescope Facility (SIRTF, 1998)

In addition, Japanese and Soviet spacecraft will complement these missions. For example, the Japanese VSOP and Soviet Radioastron satellites will study the radio signals emitted by the mysterious and powerful quasars. Together, all these missions should help unravel the mystery of the missing matter and perhaps provide critical clues for scientists trying to unify the laws of physics.

Our exploration of the cosmos will probably one day find the biggest prize of all: proof of an alien civilization. In just the last two years, there have been important developments in that search. A team of scientists led by David Latham of the Smithsonian Astrophysical Observatory made the first independently verified detection of planets around another star. There have been many alleged detections of such planets before, but they had never been confirmed by other astronomers. This breakthrough, plus work by Bruce Campbell of the University of Victoria, British Columbia, and other astronomers, strongly suggests that planets are common—*not* rare—in the universe, making it likely that life is common, too.

The Planetary Society and Argentina reached an agreement in which Argentine researchers will duplicate the Society's Harvard project, currently the world's most powerful search for extraterrestrial intelligence (SETI), looking for radio signals from other civilizations. This will be the first full-time SETI program in the southern hemisphere, where the sky has rarely been searched for this kind of signal. And NASA, the SETI Institute, Stanford University, and a new Silicon Valley company called Silicon Engines have together produced a powerful SETI chip designed to rapidly process signals from space. One day, I expect, you'll be able to buy such a chip from Radio Shack to use on your satellite dish-antenna.

When the NASA system is operational, probably by 1992, it will become the most powerful SETI program in the world. In the meantime, other researchers in the U.S., Canada, Argentina, and the Soviet Union are progressing in their own searches. The Soviets are also planning to orbit a radiotelescope soon, which may be used occasionally for SETI. Diplomats and scientists are even anticipating the first contact by discussing what we should do if that happens. They are preparing a SETI Protocol, a document in which international bodies would agree on how to handle the contact—our first example of interstellar diplomacy.

In many ways, what we're learning about the universe is revealing more about our own planet. Comparisons of Mars and Venus with Earth are teaching us much about our own weather, climate, and geology. The year 1988 saw a remarkable growth in the awareness that Earth is not the center of the universe. *Time* magazine named Earth "Planet of the Year." (Is Mars next?) And AT&T began broadcasting videotapes of people's greetings into space. Ted Koppel hosted a broadcast called "News from Earth," an ABC television special officially intended for any galactic community that might someday receive the signal. The space age is seeping into the public consciousness.

Yet a national survey the same year by the Public Opinion Laboratory of Northern Illinois University found that 21 percent of Americans think the Sun moves around the Earth, and an additional 7 percent weren't sure. It's only been 500 years since Copernicus showed that it's the other way around. Some people just haven't gotten the news yet.

Earth is the target of many spacecraft to be launched. In fact, the world's largest cooperative space project ever is called Mission-to-Earth. In 1992, the U.S., the Soviet Union, the European Space Agency, Japan, China, Australia, Brazil, Pakistan, and others will focus on our planet in an attempt to understand what's happening with the greenhouse effect that seems to be warming the place up, the ozone layer's holes that threaten to let in more dangerous ultraviolet sunlight, the destruction of forests in the Amazon and elsewhere, and the pollution that endangers our water and air.

NASA and France have been working for years on a project that should be an important part of that mission. Called TOPEX (Topography Experiment) by NASA and Poseidon by the French, it will use radar to measure the height of the oceans more precisely then ever before (to within 5 inches) and will chart the oceanic currents.

NASA is also building an Earth Observing System (EOS) for 1995 designed to study the whole planet systematically, one goal being to be able to predict changes. It will probably use two huge satellites, around 60 feet long, with instruments systematically studying our planet's atmosphere, oceans, ice bodies, land, and interior. Never before has there been such a unified effort to coordinate so many types of instruments, "like some big orchestra, the brass section talking to the woodwinds," as one scientist described it. "The idea is to observe on a time-scale longer than the time-scale of a presidential administration."

To me, the most exciting new Earth-oriented satellite is a Canadian project. Called Radarsat, it will be launched in 1994 into a polar orbit on a rocket provided by the U.S. It will build on the synthetic-aperture radar technology that allows a small antenna to act as if it were miles long. This type of radar has already discovered ancient Mayan ruins in Central America and long-lost riverbeds under the Sahara desert. It will be used to study the oceans and polar ice, to search for natural resources, and will even have applications for agriculture. One of the most economically important crops in the world is rapeseed, used for vegetable oil, and, amazingly, such radar can detect the growth of the crop even before optical satellites see it.

While the importance of outer space to our future often escapes our leaders in Washington, California congressman George E. Brown, Jr. was able to insert a remarkable Space Settlements section into U.S. Public Law 100-685, the NASA appropriations bill. It says in part, "The Congress declares that the extension of human life beyond Earth's atmosphere, leading ultimately to the establishment of space settlements, will fulfill the purposes of advancing science, exploration, and development and will enhance the general welfare." It requires NASA to periodically report on technology for space colonies (while encouraging international and private approaches), use of extraterrestrial resources, and progress toward human outposts on the Moon and an expedition to Mars.

Research is taking place that will profoundly reduce the cost of getting into space. For example, at the Jet Propulsion Laboratory (JPL) and the Autonomous University of Barcelona, mathematicians Edward Belbruno and Jaume Llibre have made breakthroughs in the calculation of orbits and trajectories that could greatly reduce the amount of fuel needed to get to the other planets. This may bring the cost of interplanetary exploration down to the level at which the European Space Agency, Japan, China, and other countries could compete with the superpowers. Already, Israel has launched its first satellite from within its own borders, and Brazil is preparing to do so. Soon, almost any nation or company that wants to put an object into space will be able to do so.

Also at JPL, engineer Ross Jones has been pushing the design of microspacecraft, grapefruit-sized probes that could be mass-produced cheaply and launched by electromagnetic guns to explore the planets. MIT's Rodney Brooks has built intelligent six-legged, insectoid robots the size of puppies that may one day explore Mars. JPL's James Burke has patented an amazing microspacecraft that could be shot to Mars, flutter down like a maple seed, and send pictures back. One day, we may seed the planets of our solar system with hundreds of tiny robots and watch as they float gently down like autumn leaves.

The overall news of the last two years has been upbeat. Science fiction is becoming reality ever faster. Researchers are making great strides in using lasers to propel rockets—thus avoiding the standard rocket problem of having to bring all your fuel along, so that the fuel not only has to accelerate the payload, it has to accelerate itself. Within just five years, the Lawrence Livermore lab could launch such a laser-propelled spacecraft. Also, solar sailing, using the pressure of sunlight, and ion drives, using electrically charged particles, should become available soon to allow cheap, fast flights for space probes to the other planets.

And in another area that borders on science fiction—antimatter—developments are happening so rapidly that scientist Robert L. Forward now publishes a newsletter on the subject. Antimatter—matter that annihilates ordinary matter, releasing 100 percent of the energy that Einstein told us was locked within it—is now being produced and stored in the world's particle accelerators. Today, it's in microscopic amounts, but tomorrow, there could be enough to make practical fast trips throughout the solar system. Its potential is so great that the RAND Corporation think tank now officially recommends a $400 million national program aimed at the practical application of antimatter, such as propulsion.

Several private American companies are building rockets with little or no governmental funding, in an attempt to break away from the often stifling bureaucracy that plagues both the civilian and military governmental space programs and adds enormously to the real cost of space flight. A study by Gary Hudson and Michael Hyson of Pacific American Launch Systems contends that we could put a twelve-person outpost on the Moon for less than $1 billion, and could rotate the crew and resupply them every three months for $100 million a year. These figures are far smaller than the usual governmental studies compute.

With more competition in the space marketplace domestically and abroad, and with many innovative designs about to be tested, the next decade should witness a drastic reduction in the cost of getting cargo into space. The decade after that, it could become relatively inexpensive for humans to go there. Orbiting Hiltons, Holiday Inns, and Howard Johnsons could become a reality.

Meanwhile, back at planet Earth, people who complain about the money "wasted" on space exploration awaken to front-page stories on global pollution, the greenhouse effect, and the hole in the ozone layer. Full understanding of each of these requires Earth-oriented spacecraft, and benefits profoundly from comparison with the weather and climate of other planets. While these people are dragged kicking and screaming into the space age, those of us who are already there will nod sagely while we explore the solar system, and watch as the Hubble Space Telescope takes snapshots of the birth of the universe.

ONE

From Fire Arrows to Neil Armstrong

THE HISTORY OF SPACE EXPLORATION

When the Chinese launched the first rocket a thousand years ago, they could not have foreseen that one day their "fire arrows" would be able to circle the Earth and travel to other worlds.

Our space heritage began with the invention of gunpowder in China, even before the invention of guns. The Chinese discovered that a mixture of saltpeter, sulfur, and other chemicals would burn fiercely, and they first used it to make firecrackers by stuffing it into bamboo. Eventually someone tried tying one of these firecrackers onto the end of an arrow, and before long the arrows traveled like simple rockets, much farther than is possible from a bow. The military used these to terrify such enemies as the Mongol invaders.

Through the Arab and Persian trade network, word of this substance—gunpowder—traveled to the Western world, and eventually the substance was manufactured in Europe.

There, it was Britain's William Congreve—a 19th-century version of the Germans' weapons mastermind, Wernher von Braun—who gave the greatest boost to this technology. Congreve, the son of the Comptroller of the Royal Laboratory at Woolwich, realized that the rocket was potentially a devastating weapon of war, and he worked hard to invent new ways of using it and making it more effective. By the time of the War of 1812, the second armed conflict between Britain and America, he had already seen the use of his weapon in Europe, and he had converted the ship *Erebus* into a "rocket ship." (This was the ship that fired the rockets that Francis Scott Key saw in the sky during the bombardment of Fort McHenry near Baltimore, inspiring him to capture that image—"the rockets' red glare"—in a poem that eventually became the U.S. national anthem.) Some of Congreve's rockets had explosives packed in them, anticipating the modern intercontinental ballistic missile.

These 19th-century rockets stimulated Jules Verne to speculate about the possibility of using them to get us off this planet. Prior to Verne, most philosophers and writers who had thought about travel to other worlds envisioned methods that can only be called pure fantasy, such as harnessing the power of dewdrops evaporating into the air, or flying into space on wings. But Verne envisioned something extraordinarily close to what did eventually happen in the 20th century. In his book

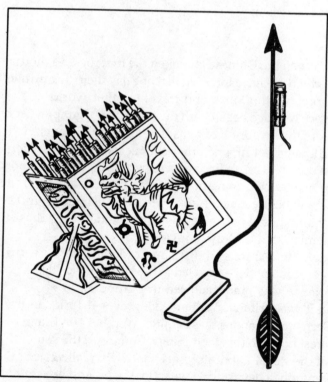

Ancient Chinese fire-arrow and launcher.

Those who dream by night in the dusty recesses of their minds wake in the day to find that all was vanity; but the dreamers of the day are dangerous men, for they may act their dream with open eyes, and make it possible.

—*T.E. Lawrence*, Lawrence of Arabia

From the Earth to the Moon, he described a giant cannon in the ground, using powerful explosives at a location in Florida. He would have been pleased to learn that in 1969 a real rocketship, launched from Florida, successfully orbited the Moon as he had predicted.

Jules Verne knew science and engineering well enough to know a rocket would not be able to land on the Moon without some other propulsion system. So, in his description, the rocket flew around the Moon and returned without having landed. In fact, that is similar to what we did with the *Apollo 8* mission, half a year before the actual Moon landing.

Another 19th-century discovery greatly influenced both science fiction and the future space program: An Italian astronomer, Giovanni Schiaparelli, studied Mars and found strange markings on the surface that he described in Italian as *canali*—channels or grooves, such as those that might be created by naturally flowing water. In English, the word was mistranslated as "canals," which are, of course, artificial.

Percival Lowell read about these canals and became obsessed with the idea that they were signs of an advanced extraterrestrial civilization on Mars. He devoted his life to studying that planet and became the greatest Mars astronomer in the world. He built an observatory, an institution that still stands today, in the Arizona mountains, primarily to watch that one planet.

Lowell wrote numerous books about the Martian civilization, and the public loved them. They, too, became fascinated with the idea of life on Mars, a concept that inspired generations of scientists and science-fiction writers. Lowell's speculations eventually spawned such diverse creations as H. G. Wells's *War of the Worlds,* and the real *Viking* spacecraft that landed on Mars in 1976. Wells's novel showed the Martians invading the Earth and trying to conquer it, and, in a sense, the *Viking* landers were our answer to that challenge—robot laboratories designed specifically to look for life on the surface of Mars.

Also in the 19th century was born the Russian founder of the modern science of astronautics, Konstantin Tsiolkovsky. A largely self-educated scientist, he developed the theory of space travel and, through his writings, both scientific and science-fictional, inspired generations of Russians to think about space exploration and to work toward some-day making his dreams of space colonization a reality. He lived through the Russian Revolution and was honored afterward. His books and scientific papers are continuously in print in the Soviet Union, serving as an inspiration to future Russian scientists and engineers.

Fiction into Fact

In the early decades of the 20th century, two scientists built on Tsiolkovsky's inspiration and finished the groundwork for the development of modern rockets. One was Transylvania-born Hermann Oberth and the other an American, Robert Goddard. In 1929, Oberth served as a technical adviser to the German director Fritz Lang on the movie *Die Frau im Mond,* released in English as *The Girl in the Moon.* This film inspired a young German named Wernher von Braun to start thinking about space travel—and about more earthly uses for rockets.

What, technically, is a rocket? Basically, it's a device that works by pushing a fuel out to cause it to move. The most primitive rocket is a toy balloon filled with air and then released. The air shoots out the opening in the balloon, pushing it away. One of Isaac Newton's laws of motion states that for every reaction there's an equal and opposite reaction. What this means here is that as the molecules of air flow out of the balloon to the right, they push the balloon to the left.

This is why *The New York Times* was wrong back in 1914 when it said Goddard didn't know basic physics. The editors thought there was no way that a rocket could work in space, because there was nothing for it to push against. But in a rocket, the fuel pushes against the vessel, so it doesn't matter whether there is air or not. So it was the editors of *The New York Times* who didn't know basic physics. And, to their credit, decades later, when spacecraft had orbited the Earth, the *Times* apologized in an editorial to Goddard.

Nowadays, we generally distinguish between a jet and a rocket. A jet is an "air-breathing" aircraft—a special rocket, one that carries only fuel, not oxidizer. It uses air, the oxidizer that allows the fuel to burn.

The main thing, it seems to me, is to remember that technology manufactures not gadgets, but social change. Once the first tool was picked up and used, that was the end of cyclical anything. The tool made a new world, the next one changed that world, the one after that changed it again, and so on. Each time the change was permanent. Using the tool changes the user permanently, whether we like it or not. Once when I was in Moscow talking to academician Petrov, I said, "Why don't you buy American computers to get you into space quicker and more effectively?" He replied, "No fear; they'd make us think like Americans."

—James Burke, author and television journalist

A true rocket, on the other hand, carries its own oxidizer, so it doesn't need air to operate. The Space Shuttle, for example, carries tanks of liquid hydrogen and liquid oxygen. The hydrogen is the fuel; the oxygen is the oxidizer that burns the hydrogen, producing the great flames spewing from the three main engines. It needs to carry along oxygen because it leaves the dense part of the atmosphere so quickly, reaching the upper atmosphere where oxygen is rare.

The Space Shuttle also uses two solid rocket boosters, one of which caused the Space Shuttle *Challenger* to explode. Instead of using liquid fuel, like the main engines of the Shuttle, the solid rocket boosters—like Chinese fire-arrows—contain a fuel that is more or less a mixture of rubber, salt (ammonium perchlorate), and aluminum, serving as fuel, oxidizer, and stabilizer, respectively.

Dreams into Nightmares

Before World War II, Hermann Oberth proposed to the German War Department the development of a liquid-fueled, long-range missile. (After the war, he recommended that similar rockets be used in interplanetary space flight.) At about the same time Oberth began to develop his ideas, Goddard was doing similar work in the U.S. But where Oberth was primarily a theoretician working with government institutions, Goddard was an experimentalist on his own. He built actual rockets and tried them out. Many of them failed, but some did travel high into the sky. His work was widely ignored by scientists, engineers, and military officers in the U.S., but was carefully studied in Germany, and strongly influenced their World War II rocket program.

Though World War II in many ways fulfilled the dreams of Tsiolkovsky, Oberth, and Goddard, the dreams turned into a nightmare.

The German scientist Wernher von Braun built upon ideas he'd studied as a boy, shooting off homemade rockets. With the arrival of World War II, he began supervising a large part of the Nazi rocket research enterprise. The Germans concentrated on building two types of missiles, the V-1 and the V-2, both unpiloted. The V-1 was a pulse-jet-propelled airplane, very much like the modern cruise-missile. Launched from the continent, it traveled on airplane-like wings across the English Channel and struck Britain, where it terrorized the citizens. It traveled slower than sound, so its roar could be heard before it hit, giving people a chance to take cover, and earning it the nickname of the "buzz-bomb."

Wernher von Braun worked on the true rocket, the V-2, an outgrowth of Goddard's experiments. This was the one von Braun believed would someday lead to a rocket that could take him into space. The V-2 was the forerunner of today's intercontinental ballistic missile (ICBM). It took off in a vertical position, like one of today's rockets, and did not have wings, except for small tail fins to control the rocket's direction. It roared off the launch pad and accelerated until it traveled faster than the speed of sound, so it outraced its own noise. The English (and Dutch) who were hit had no warning, making it an even more terrifying weapon than the buzz-bomb. You could be happily tending your garden one moment, only to find your world suddenly exploding around you a split-second later.

In 1942, the first successful launch of the V-2 rocket occurred at the Peenemünde launch site on the Baltic Sea. The commander of Peenemünde, Major General Doctor Walter R. Dornberger, said, "Today the spaceship is born."

The V-2 rocket was never very successful as a weapon of war. It was more complex than the V-1. It took much longer to develop and so was not used effectively by the time the war came to an end.

Not many people realized it, but the Germans had jet airplanes and

> *It seemed likely that, if the German had succeeded in perfecting and using these new [rocket] weapons six months earlier than he did, our invasion of Europe would have proved exceedingly difficult, perhaps impossible.*
>
> —Gen. Dwight Eisenhower

rocket planes during World War II. These were piloted aircraft flown against the Allies. The Messerschmitt 163B V18 Komet, for example, reached an incredible 702 miles per hour on July 6, 1944.

The rocket fighter was so dangerous that it wound up killing more German pilots than enemies. It's chilling to think that had the Germans developed their rockets and jets earlier in the war and perfected them, they might have been able to wipe out the Allied air forces in the sky during World War II.

The main Allied contribution to this technology during the war was the development of JATO (Jet-Assisted Take-Off) units, developed by the American military along with a Caltech lab (which later became the Jet Propulsion Laboratory, JPL). The JATO units were rockets strapped onto the bodies of airplanes to allow them to take off from short runways. (In fact, the Jet Propulsion Laboratory never had anything to do with turbojet engines; it always worked with rockets. But rockets were thought too Buck-Rogerish to be taken seriously.)

When the war was coming to an end and it became clear that the Nazis were losing, most of the German rocket scientists joined von Braun, who took them on a journey to surrender to the Americans. With forged German orders, they traveled through the crumbling Nazi lines, using trains and a hundred cars and trucks filled with equipment and research records.

They were captured and held hostage by the Gestapo, escaped, and waited for General Patton's army to arrive. But the Americans had run out of gas, so the German scientists sent Magnus, von Braun's English-speaking brother, to meet them. The German scientists showed the Americans where they had sealed tons of documents in a Bavarian cave to prevent their being destroyed by the Nazis. The Americans then smuggled the scientists overseas, first to Texas, and then to White Sands Proving Grounds in New Mexico. During the years following, the German scientists reassembled V-2s from the captured equipment and launched them.

In 1949, the Germans walked over the border at El Paso, Texas, into Mexico, then walked back across into the States, where they were given official immigrant status.

A few German scientists had sought refuge with the Russians, while others were captured by the Soviets as the war came to an end, so both sides wound up with German rocket expertise. This led to the postwar joke comparing the Russian and American space programs: "It's all just a matter of whether our Germans are better than their Germans."

Footsteps into the Universe

It was there, in White Sands, that the next major step into space was achieved: the use of the multistage rocket. This got around the problem of a rocket fuel not being powerful enough to lift its own weight off the Earth into space.

One pound of any rocket fuel, no matter how powerful, does not have enough energy to lift that pound from the surface of the Earth into orbit. And there is not enough energy in any rocket fuel to give that fuel the speed it needs to go into orbit. So, at first glance, it seemed to scientists that trying to get a rocket into orbit would be futile. But throughout the history of technology, we have found that when something looks impossible, we just have to be more clever.

This "new" approach had actually been thought of hundreds of years ago. It meant breaking the rocket into two or more pieces. The simplest solution would be a two-stage rocket, with one big rocket on the bottom and a little one on top. The two rockets would fly together, with only the large one burning, until it ran out of fuel. At that point, the two rockets would separate, and the big rocket, now just a heavy, empty shell, would fall back to Earth, while the little one started to burn. This smaller craft, having received enormous momentum from the larger one, could now achieve much higher speeds and altitudes than it could have on its own. A combination of three or more stages could enable it to leave the planet completely.

I was a kid at the White Sands Proving Grounds (now known as White Sands Missile Range) in the 1950's; my father, an Army officer, helped fire these experimental rockets. Back then, the world's altitude

Though the crystal ball is cloudy, two things seem clear:

1. A satellite vehicle with appropriate instrumentation can be expected to be one of the most potent scientific tools of the Twentieth Century.

2. The achievement of a satellite craft by the United States would inflame the imagination of mankind, and would probably produce repercussions in the world comparable to the explosion of the atomic bomb.

Such a vehicle will undoubtedly prove to be of great military value. However, the present study was centered around a vehicle to be used in obtaining much-desired scientific information on cosmic rays, gravitation, geophysics, terrestrial magnetism, astronomy, meteorology, and properties of the upper atmosphere. . . .

The most fascinating aspect of successfully launching a satellite would be the pulse-quickening stimulation it would give to considerations of interplanetary travel.

—Francis H. Clauser, David T. Griggs, and Louis Ridenour,
Preliminary Design of an Experimental World-Circling Spaceship,
Project RAND Report, May 2, 1946

record was achieved by a V-2 rocket with a small WAC Corporal rocket on top of it.

This two-stage combination achieved an altitude of 250 miles, well above the 100-mile point at which space is said to begin. It was still a long way from being a true space shot because the rocket simply went up and came down again, like a baseball thrown 250 miles into the air.

The space program was following a pattern suggested 300 years earlier by Isaac Newton. Newton envisioned what modern scientists call a "thought experiment." To explain the law of gravity, he imagined a tall mountain, with a cannon on top. Picture a cannonball fired horizontally, said Newton. The cannonball travels out and falls to the Earth, perhaps a mile away. But if you made the cannon more powerful, you could shoot the cannonball farther, and it might land hundreds of miles away. If you could shoot it with even more powerful explosives, it might travel thousands of miles.

The Earth is only 8,000 miles in diameter. If you had a fantastically powerful cannon, you could shoot that cannonball so far that it would travel parallel to the surface of the Earth until it came all the way back to where it started—and hit you! The cannonball would have orbited the Earth. In practice, of course, air resistance would prevent this from working. But if you got high enough, or if the air were not there, this would be a way of putting an object into orbit around the planet.

Sputnik: The Shot Heard Round the World

When I was at White Sands, we kids would sometimes get out of classes for a few minutes to go outside and watch a rocket launch. In the mid-'50's, the idea of rockets going into space was widely regarded as "Buck Rogers nonsense," but we White Sands kids knew otherwise. Our parents were routinely shooting rockets into the edge of space.

So we were not as surprised as most people on that day, October 4, 1957, when we woke to hear that the Russians had launched the first spacecraft, and that it was orbiting the Earth. Their German scientists had beaten ours after all, just one month short of the 40th anniversary of the Russian Revolution. The U.S. probably could have launched a satellite before the Russians, if it had had a high priority, but history would record that the first artificial moon had a Russian name: *Sputnik,* meaning "fellow traveler."

Sputnik shocked the West. The Russians, who had seemed to us so technologically backward, were suddenly in the forefront of modern science. What followed, of course, was a fierce scramble in the U.S. to launch a satellite of our own into orbit. The Space Age had become a space race.

Behind the scenes, competition raged between the Army, the Navy, and the Air Force, sometimes to the point of comical subterfuge. At one point, it is rumored, scientists actually hid a satellite in a closet so that government inspectors wouldn't detect that they were doing unauthorized work that was supposed to be done by the Air Force.

The month after the first *Sputnik,* the Russians launched the first animal into space, the dog Laika. Then we launched our first *Vanguard* rocket (a civilian/Navy product), and it blew up.

Finally, in the beginning of 1958, *Explorer 1* (from the Army) finally orbited the Earth, and America was in the space business.

The U.S. tried hard to catch up to the Russians. But less than four years after *Sputnik,* the first human being to orbit Earth bore a Rus-

NATIONAL AERONAUTICS AND SPACE ACT OF 1958,
AS AMENDED
Sec. 102.

(a) The Congress hereby declares that it is the policy of
the United States that activities in space should be devoted
to peaceful purposes for the benefit of all mankind.

(b) The Congress declares that the general welfare and
security of the United States require that adequate provision
be made for aeronautical and space activities. The Congress
further declares that such activities shall be the responsibility of, and shall be directed by, a civilian agency exercising
control over aeronautical and space activities sponsored by
the United States, except that activities peculiar to or
primarily associated with the development of weapons
systems, military operations, or the defense of the United
States (including the research and development necessary to
make effective provision for the defense of the United
States) shall be the responsibility of, and shall be directed
by, the Department of Defense. . . .

(c) The aeronautical and space activities of the United
States shall be conducted so as to contribute materially to
one or more of the following objects:

(1) The expansion of human knowledge of phenomena
in the atmosphere and space;

(2) The improvement of the usefulness, performance,
speed, safety, and efficiency of aeronautical and space
vehicles;

(3) The development and operation of vehicles capable
of carrying instruments, equipment, supplies, and living
organisms through space;

(4) The establishment of long-range studies of the potential benefits to be gained from, the opportunities for, and
the problems involved in the utilization of aeronautical and
space activities for peaceful and scientific purposes;

(5) The preservation of the role of the United States as a
leader in aeronautical and space science and technology and
in the application thereof to the conduct of peaceful activities within and outside the atmosphere;

(6) The making available to agencies directly concerned
with national defense of discoveries that have military value
or significance, and the furnishing by such agencies, to the

> *civilian agency established to direct and control nonmilitary aeronautical and space activities, of information as to discoveries which have value or significance to that agency;*
>
> *(7) Cooperation by the United States with other nations and groups of nations in work done pursuant to this Act and in the peaceful application of the results thereof; and*
>
> *(8) The most effective utilization of the scientific and engineering resources of the United States, with close cooperation among all interested agencies of the United States in order to avoid unnecessary duplication of effort, facilities, and equipment. . . .*
>
> *(e) . . . development of advanced automobile propulsion systems. . . .*
>
> *(f) . . . programs designed to alleviate and minimize the effects of disability. . . .*

sian name—Yuri Gagarin. Buck Rogers was no longer nonsense. So President Kennedy, stung by this defeat and by the disastrous Bay of Pigs invasion of Cuba, announced the objective of putting a man on the Moon by the end of the decade.

The space race had a new destination.

The race to the Moon had actually started earlier with a great thud. The Americans launched the first unmanned lunar probe, called *Able 1*, but it failed, exploding one minute after launch.

That was in August 1958. In January 1959, the Russians successfully flew *Luna 1* past the Moon. Then they actually impacted a probe, *Luna 2*, in September of that year. And the next month, *Luna 3* returned the first pictures ever taken of the far side of the Moon. So great was the humiliation in America because of the Russian space successes that an article was published claiming that the Russian Moon pictures had been faked. In fact, it did turn out that the photographs had been retouched by hand since the Russians lacked the advanced computer processing that became a hallmark of American space photography.

In March 1959, our *Pioneer 4* became the first American spacecraft to fly by the Moon, passing it at 40,000 miles. This was one of a series of disastrous American space missions aimed at the Moon, most of them called *Ranger*.

The *Rangers* were designed to crash-land on the Moon, hitting it without slowing down, taking pictures as they went in on their kamikaze missions, postponing to future missions the difficulty of landing softly. To people soured on the space program because of its recent disasters, I suggest a look at the roll call of the *Rangers: Ranger 1* had a rocket malfunction. *Ranger 2*'s rocket also malfunctioned. *Ranger 3* missed the Moon by 20,000 miles. *Ranger 4* had both instrument and guidance failure, and most likely landed on the far side of the Moon. *Ranger 5* missed by 400 miles. *Ranger 6* hit the Moon but sent back no pictures. Only after a great shake-up in the program at JPL did success finally come with *Ranger 7* in 1964. At last, NASA was getting great close-up pictures of the Moon.

And while all those early *Ranger* missions were missing their target, the Russians' luck seemed to change as well. Though the first three lunar probes the Soviets had sent to the Moon were remarkably successful, the later ones were disasters. *Luna 4* missed the Moon by 5,000 miles. *Luna 5* crash-landed on the Moon. *Luna 6* missed the Moon by 100 miles.

The three men most responsible for the first successful American spacecraft, *Explorer 1:* William H. Pickering (left), JPL director; James Van Allen (center), University of Iowa scientist whose instrument detected the radiation belts around the Earth named after him; and Wernher von Braun, leader of the Army team that built the first stage of Redstone rocket used.

Finally the Russians got back on track with *Zond 3* in 1965, which again photographed the far side of the Moon. From then on, they had a mixture of successes and failures in their lunar probes, all of them unmanned.

After *Rangers 7, 8,* and *9* had proved overwhelmingly successful, the American *Surveyor* program soft-landed spacecraft on the Moon, most of them successfully, starting with *Surveyor 1* in 1966. These spacecraft landed gently by using additional rockets to slow them down near the Moon. They carried cameras to take close-up pictures, and probes to measure the properties of the lunar surface.

Also in the early sixties, scientists began to turn their eyes toward the planets, particularly the two nearest: Venus, just inside Earth's orbit, and Mars, just outside. In 1961, two months before Yuri Gagarin went into space for the first time, the Russians launched the first interplanetary probe, *Venera* (Venus) *1,* toward the planet Venus, but it died before it got there. The next year, the U.S. launched *Mariner 1* toward Venus; this effort failed as well. However, one month later, *Mariner 2* was launched, and it succeeded, the first spacecraft ever to fly by another planet.

The next year, the Russians launched *Zond 1* to Venus, but it, too, was unsuccessful. The U.S. *Mariner 3* to Mars also failed. However, *Mariner 4,* launched the same year, became the first probe to fly by Mars. The Russians continued with a series of probes to Mars that were mostly unsuccessful, while the American probes to that planet became increasingly complex and reliable.

In 1965, the same year that *Mariner 4* successfully flew by Mars, the Russians hit Venus with *Venera 3.* Although the probe itself failed to yield any information, it was the first object from Earth to hit another planet, and so, despite its failure, marked another milestone in our movement into the universe.

From this point on—until our recent tragedies—both the Russian and American space programs really took off, with only a few setbacks. In 1967, the Soviet *Venera 4* became the first probe to enter the Venusian atmosphere successfully, returning data as it plunged through the hot, thick air. Then in 1970, they succeeded in landing a spacecraft, *Venera 7,* on the surface of that infernal planet. It transmitted data for 23 minutes from the surface before dying from the heat. This was the first successful soft-landing of an object from the planet Earth onto another planet.

On March 2, 1971, the U.S. launched *Pioneer 10,* a spacecraft pro-

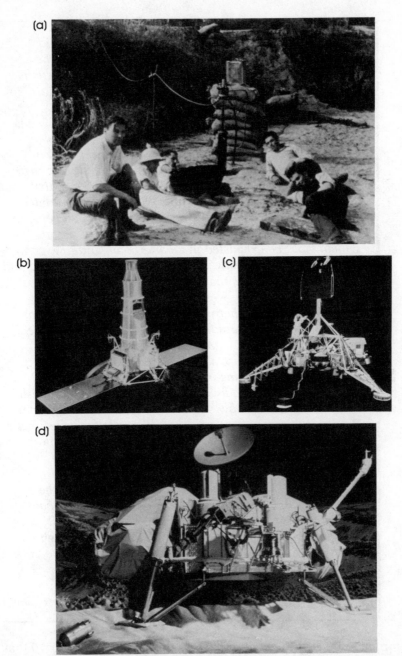

(a) First rocket test at the place that became the Jet Propulsion Laboratory.
(b) *Ranger 7*, the first U.S. spacecraft to reach the Moon.
(c) *Surveyor*, the first spacecraft to soft-land on the Moon.
(d) Our first robot built for Mars: the *Viking* lander.

We never gain as much from [exploration] as the wild enthusiasts promise; we invariably gain more than the frightened old men predict. And regardless of predictions, the exploration must go on because it is in man's nature to explore. . . .

When I was a little boy in a small town in Pennsylvania, past my door ran a remarkable road. To the east it went a quarter of a mile and stopped dead. To the west it was limitless. It went all the way to the Pacific, and from there to Asia and the entire world. As a child I looked at that road and understood its two directions—limited and unlimited—and thought how craven it would be for a human being to devote his life to the exploration of the eastern portion, which could be exhausted in an afternoon, and how commendable to turn westward and thus enter upon a road and a complexity of roads that would lead to the very ends of the Earth. I chose the western road. . . .

I think, however, that when one deals with exploration, one has got to be aware that in every generation one field of exploration ends. . . . As that epoch ends, we start something new. We are always at the end of something, always at the beginning of societies, not only of total culture, but also of individuals. If we have no accomplishment, if we never know success, we lead embittered lives. But if we stop with one success and do not recognize that it stands merely as a threshold to something greater, more complex, more infinite, then I think we do only half our job. . . .

It is this kind of threshold that has always made the explorer's life exciting. . . . I am . . . apprehensive about the explorations, yet absolutely certain that they will go forward and that the triumphs and defeats that go with them will form a basic characteristic of man and one of the best characteristics.

—*James Michener, author*

grammed to fly by Jupiter. So great was the speed of *Pioneer 10* that after sending back its observations of Jupiter, it flew beyond the limits of our solar system, becoming the first interstellar spacecraft.

Pioneer 10 carried a plaque, designed by Carl Sagan and colleagues, showing a picture of two people, a man and a woman, and giving a map of the solar system that showed, in a mathematical code, our position in the Milky Way galaxy.

Pioneer 10 also marked the Americanization of the outer solar system. No Soviet probes have been launched toward any planet beyond Mars, while four American probes have. *Pioneers 10* and *11* were flybys of Jupiter and Saturn, followed by *Voyagers 1* and *2*, much more sophisticated robotic explorers that took extraordinarily high-quality pictures of those two planets. The *Voyagers* were originally intended to be part of a Grand Tour of the outer solar system. NASA had hoped to send spacecraft to visit all the outer planets: Jupiter, Saturn, Uranus, Neptune, and Pluto. However, lack of funds forced a cutback of this program, and it was scaled down until it became the *Voyager* mission to Jupiter and Saturn. Nevertheless, both spacecraft were built so well that neither the intense radiation belts of Jupiter nor the particles in the rings of Saturn were able to destroy them, and *Voyager 2* was put into a path toward Uranus.

In 1986, that spacecraft survived triumphantly to fly by Uranus and was put into a trajectory that would take it to the planet Neptune. The old Grand Tour is to be accomplished, even with the cut-rate spacecraft: all the outer planets visited except Pluto.

Buck Rogers for Real

One month after Gagarin orbited the Earth, Alan Shepard saved some face for the West by flying some 300 miles into space, making a suborbital flight. It wasn't until the next flight, on February 20, 1962, that John Glenn became the first American to orbit the Earth, after the Soviets had orbited their second cosmonaut, Gherman Titov.

The Russians continued to set the pace by achieving the first flight by a female cosmonaut, the first three-man flight, and the first space walk. Americans outgrew their one-man *Mercury* space program and entered the next stage, the *Gemini* program, with a two-man spacecraft.

In 1966, with President Kennedy's deadline less than four years away, a *Gemini* flight nearly ended in disaster when the spacecraft began to spin out of control because of a jammed thruster. The astronauts, however, managed to solve the problems and return to Earth safely. (Both of those astronauts would later land on the Moon. One of them was Neil Armstrong.)

We choose to go to the moon. We choose to go to the moon in this decade, and do the other things, not because they are easy but because they are hard; because that goal will serve to organize and measure the best of our energies and skills; because that challenge is one that we are willing to accept, one we are unwilling to postpone, and one which we intend to win—and the others, too.

It is for these reasons that I regard the decision last year to shift our efforts in space from low to high gear as among the most important decisions that will be made during my incumbency in the office of the Presidency. . . .

The growth of our science and education will be enriched by new knowledge of our universe and environment, by new techniques of learning and mapping and observation, by new tools and computers for industry, medicine, the home as well as the school.

—President John F. Kennedy, September 12, 1962

Meanwhile, both the Americans and the Russians conducted experiments to develop docking spacecraft of the type needed for lunar missions.

Just when it seemed that the American program finally was rocketing to success, disaster struck. In January 1967, the first *Apollo* spacecraft, sitting on the ground for a routine test with three astronauts, caught fire. Astronauts Virgil "Gus" Grissom, Edward White, and Roger Chaffee died. Our goal of reaching the Moon before the end of the decade looked impossible.

Nevertheless, NASA persisted, redesigning the spacecraft, and from then on the *Apollo* program had a series of successes, beginning with the first piloted flight of the *Apollo* series, *Apollo 7*, and the first space flight to orbit humans around the Moon, *Apollo 8*. Astronauts Frank Borman, James Lovell, and William Anders became the first humans to travel to another world, even if they didn't land on it. They orbited the Moon at Christmas time, 1968, very much as Jules Verne had imagined in the previous century.

Then, finally, on July 20, 1969, the world watched as Neil Armstrong and Buzz Aldrin landed on the Moon, while Michael Collins orbited around it. Human beings, for the first time, set foot on another world.

After our great push toward this achievement, the Russians said they hadn't really intended to land a man on the Moon after all. People wondered if the space race had been a fiction created by NASA. But the history of the Russians' space probes makes it clear that they'd been trying mightily to plant the hammer and sickle before the stars and stripes got there.

We went back and landed on the Moon seven times in the *Apollo* program, although one of those missions, *Apollo 13,* nearly ended in tragedy when an oxygen tank exploded en route. *Apollo 13* may have had an unlucky number, but the ingenuity of the engineers on the ground and the improvising ability of the astronauts in the spacecraft allowed it to return to Earth safely, though it did not land on the Moon. (The near-disaster occurred simply because someone installed a wrong part in that oxygen tank; the lives of space-farers often hinge on such small matters.)

Following the first *Apollo* Moon landing—celebrated all over the world—the American public became bored with outer space. NASA and the aerospace industry, having accomplished one of the most incredible feats in the history of the human race, were suddenly hearing, "What have you done for us lately?" With declining public support, NASA had to cut back its plans to explore the Moon further.

The Russians, on the other hand, tend to forge ahead regardless of successes, failures, or lapses in world interest. The Russian leaders, and the average Soviet citizen, seem to believe that the exploration of space is an inevitable direction for their civilization, and they take pride in outdistancing the U.S. whenever possible.

After *Apollo*

Following our last landing on the Moon, NASA put into orbit the first American space station, *Skylab.* Crews of three astronauts were to spend one to three months in space. The station was abandoned after the third crew returned, and *Skylab* eventually fell back to Earth, landing in the Australian desert. Unfortunately, money had not been available to put *Skylab* into a higher orbit and maintain it as a permanent orbiting space station.

The *Apollo* program came to a conclusion with the *Apollo-Soyuz* mission, the first joint manned-space project between the U.S. and the Soviet Union, which occurred in July 1975, three months after the

last U.S. troops left Vietnam. Three *Apollo* astronauts linked up in orbit with two Soviet cosmonauts. They visited each other's spacecraft
and shared food. This was the high point of Soviet-American cooperation in space. Afterward, the cold war heated up again, and relations
between the two countries deteriorated—until recently, when cooperation surprisingly improved on space programs, if not in other spheres.

At the close of the *Apollo* program, NASA began concentrating on
the development of a reusable rocket, the Space Shuttle. They argued
that the airplane would never have been economical if, every time
you built a Boeing 747, you flew it once and threw it away. Why not
build a spacecraft that could be reused for many flights?

POST-APOLLO SPACE PROGRAM

*We believe the Nation's future space program possesses
potential for the following significant returns:*

*New operational space applications to improve the quality
of life on Earth.*
Non-provocative enhancement of our national security.
*Scientific and technological returns from space investments
of the past decade and expansion of our understanding
of the universe.*
*Low-cost, flexible, long-lived, highly reliable, operational
space systems with a high degree of commonality and
reusability.*
*International involvement and participation on a broad
basis.*

*Therefore, we recommend that this Nation accept the
basic goal of a balanced manned and unmanned space program conducted for the benefit of all mankind.*

*To achieve this goal, the United States should emphasize
the following program objectives:*

*Increase utilization of space capabilities for services to man,
through an expanded space applications program.*
*Enhance the defense posture of the United States and
thereby support the broader objective of peace and
security for the world through a program which exploits
space techniques for accomplishment of military
missions.*

Increase man's knowledge of the universe by conduct of a
continuing strong program of lunar and planetary
exploration, astronomy, physics, the earth and life
sciences.
Develop new systems and technology for space operations
with emphasis upon the critical factors of 1) common-
ality, 2) reusability, and 3) economy, through a program
directed initially toward development of a new space
transportation capability and space station modules
which utilize this new capability.
Promote a sense of world community through a program
which provides opportunity for broad international
participation and cooperation.

—Space Task Group (Vice President Spiro Agnew, chairman;
Robert C. Seamans, Secretary of the Air Force; Thomas O. Paine,
NASA Administrator; Lee A. Dubridge,
Presidential Science Adviser), September, 1969

NASA and the Air Force had conflicting ideas about how the Shuttle should be built. To begin with, the Air Force didn't really want it.
They were happy with old-fashioned expendable rockets. But as long
as NASA was proposing to build a reusable spacecraft, the Air Force
wanted to be sure that it was big enough to carry the military payloads
they wanted to put into space—especially, large spy satellites. This led
to elaborate horse-trading behind the scenes. The Air Force wanted
requirements imposed upon the NASA design, but did not want to
pay any of the cost. In the meantime, Congress got into the act. Disillu-
sioned with the space program in the wake of public disinterest following *Apollo,* politicians pushed to cut back NASA's spending. Such prom-
inent senators as Edward Kennedy and Walter Mondale attacked the
Space Shuttle as a waste of taxpayers' money.

Because of all these cutbacks, NASA had to aim for the cheapest
vehicle that could still meet their major goals. This meant, ironically,
that part of the spacecraft would *not* be reusable. The large external
tank of the spacecraft would be thrown away on every flight, because
it would be too high to parachute safely back to Earth. Solid rocket-
boosters were designed to be reusable, though they were to be jettisoned early in the flight, parachuting back to Earth to be recovered
from the ocean.

. . . I was in war-torn Biafra. We were in a jeep. A plane loomed behind us out of the sun and dove down on the jeep in a strafing run. We plunged into a ditch, face down in the mud. I could contemplate that even as we were pressing our faces into the muddy Earth in safety from our brothers, men found it possible to walk erect on the Moon. That evening, the war suddenly came to a halt, at least for a few hours. The word had spread through Biafra that human beings were setting foot on the Moon for the first time. Suddenly everyone had a new perspective. It didn't last long enough to cause the war to end altogether, but for a few moments at least we could contemplate the possibilities of human grandeur and meditate on our station in infinity. In that sense, the most significant achievement of that lunar voyage was not that man set foot on the Moon, but that he set eye on the Earth.

—*Norman Cousins, author*

In 1988, the Soviet Union launched its own space shuttle, very similar to NASA's. Until recently, the Soviets were extremely secretive about their space missions and did not announce their plans in advance. Happily, that policy has changed in the last couple of years, at least for unmanned missions. They have announced plans for several space probes and have even invited American participation in such missions as their *VEGA* spacecraft to Halley's Comet.

We've come a long way since Wernher von Braun began launching tiny rockets in the 1930's. The great engineer died in 1977, never having achieved his dream of personally going into space. But he did live to see *Apollo* astronauts land on the Moon. Probably more than any other person alive at the time, he was responsible for the success of that mission. And he laid the foundation for the many successes of the American space program—some, we hope, yet to come.

TWO

Orbital Truck

THE SPACE SHUTTLE

"I expect to become an orbiting grandfather," said von Braun, just after the first *Apollo* Moon landing. He believed that, in the 1970's, he would fly a Space Shuttle that would dock at a Space Station. Although he never did, his last major gift to the space program was to help NASA settle on a Shuttle design intended to keep its cost to a minimum.

Originally, the favored design consisted of two winged rockets, one of them piggybacked on the other. The bottom one would contain the fuel to accelerate the pair to the edge of the atmosphere. Then they'd separate, the lower one flying back to Earth while the other used its own fuel and engines to get into orbit.

Von Braun and his colleagues settled on the design we use today: The lower vehicle was replaced by two solid-rocket boosters and a giant fuel tank containing liquid hydrogen and oxygen. The upper part

Whatever impedes the intercourse of the extremes with this, the center of the republic, weakens the nation. . . . Let us, then, bind the republic together with a perfect system of roads and canals. Let us conquer space.

—*John Calhoun, 1816, introducing a bill calling for Congress to issue bonds for $1.5 million*

(a)

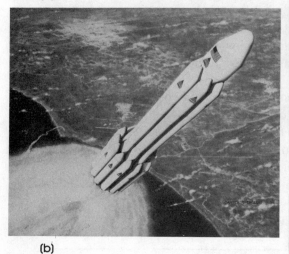

(b)

(c)

The Shuttle's competition:
(a) The European *Ariane* launch facility in Kourou, French Guiana.
(b) American Rocket Company's proposed privately financed *Industrial Launch Vehicle One*.
(c) The *Conestoga 300*, a privately financed rocket proposed by Space Services, Inc.
(d) A typical mission for the Pacific American Launch System, Inc.'s proposed privately financed *Liberty* rocket.

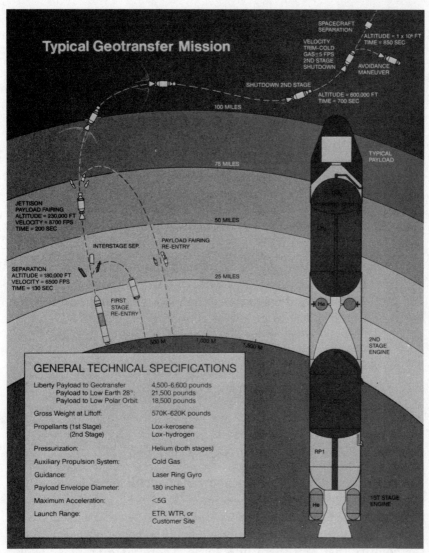

Typical Geotransfer Mission

SPACECRAFT
SEPARATION

ALTITUDE = 1 x 10⁶ FT
TIME = 850 SEC

VELOCITY
TRIM-COLD
GAS ±5 FPS
2ND STAGE
SHUTDOWN

AVOIDANCE
MANEUVER

SHUTDOWN 2ND STAGE

ALTITUDE = 600,000 FT
TIME = 700 SEC

100 MILES

TYPICAL
PAYLOAD

75 MILES

JETTISON
PAYLOAD FAIRING
ALTITUDE = 230,000 FT
VELOCITY = 8700 FPS
TIME = 200 SEC

50 MILES

INTERSTAGE SEP.

PAYLOAD FAIRING
RE-ENTRY

SEPARATION
ALTITUDE = 180,000 FT
VELOCITY = 6500 FPS
TIME = 130 SEC

25 MILES

FIRST
STAGE
RE-ENTRY

He

2ND
STAGE
ENGINE

500 M   1,000 M   1,500 M

GENERAL TECHNICAL SPECIFICATIONS

Liberty Payload to Geotransfer	4,500-6,600 pounds
Payload to Low Earth 28°:	21,500 pounds
Payload to Low Polar Orbit:	18,500 pounds
Gross Weight at Liftoff:	570K-620K pounds
Propellants (1st Stage)	Lox-kerosene
(2nd Stage)	Lox-hydrogen
Pressurization:	Helium (both stages)
Auxiliary Propulsion System:	Cold Gas
Guidance:	Laser Ring Gyro
Payload Envelope Diameter:	180 inches
Maximum Acceleration:	<5G
Launch Range:	ETR, WTR, or Customer Site

RP1

He

1ST STAGE
ENGINE

(d)

remained a winged vehicle, its three main rocket engines fueled by the external tank. On lift-off, the two solid rockets and the main engines would fire together. Two minutes after launch, the solid rockets—direct descendants of the Chinese fire-arrows—would be jettisoned, parachuting to the ocean for recovery and reuse. Six minutes later, the external tank would separate and burn up in the atmosphere, lost forever, the only part of the system not reusable.

> . . . all these possibilities, and countless others with direct and dramatic bearing on human betterment, can never be more than fractionally realized so long as every single trip from earth to orbit remains a matter of special effort and staggering expense. This is why commitment to the space shuttle program is the right next step for America to take, in moving out from our present beachhead in the sky to achieve a real working presence in space—because the space shuttle will give us routine access to space by sharply reducing costs in dollars and preparation time.
>
> —President Richard M. Nixon, 1972

The system was built, and the first fully operational Space Shuttle, the *Columbia*, blasted off with astronauts John Young and Robert Crippen in 1981. Weighing 2,000 tons at launch, the Shuttle orbited the Earth and returned with no major problems.

The Shuttle can place more than 30 tons of payload into low-Earth orbit (the spacecraft cannot go higher than 600 miles altitude), and can carry enough supplies to last a month. The complex system worked successfully through two dozen launches—until the *Challenger* launch of January 28, 1986.

The *Challenger* Disaster

On that fateful day, three days short of the 28th anniversary of the first successful American space launch, *Explorer 1*, seven astronauts took off in the Space Shuttle *Challenger*. They were a mixed group, characteristic of the diversity of people in the space program and in the country that sent them into space. There were Francis "Dick" Scobee, the experienced commander of the mission, Michael Smith, the pilot, Ellison Onizuka, Mission Specialist 1, a Japanese-American, Judith Resnik, Mission Specialist 2, an experienced astronaut/scientist, Ronald McNair, a black scientist, Mission Specialist 3, Christa McAuliffe, Payload Specialist 1, the first teacher in space, and Gregory Jarvis, Payload Specialist 2, representing industry. A cross-section of the American populace rose off the launch pad that day.

It had been unusually cold for Florida, freezing at night. There were icicles on much of the equipment on the ground. Compared to the tremendous heat of the blast-off, it seemed that such cold temperatures were irrelevant, but, alas, this turned out not to be true.

The three main engines of the Space Shuttle at the bottom of the airplane-like orbiter fired, along with the two solid-rocket boosters strapped on either side of the orbiter.

As thousands of people, including Christa McAuliffe's parents, watched from the ground, the Shuttle rose off the launch pad in a thunderous roar and a great flame. One minute later, the frozen rubber O-ring seals, separating the metal cylinders out of which the booster was made, began to burn through.

In a split second, a flame flickered through to the outside air. The flame grew until it touched the external tank. Vapor began to escape from the tanks. They exploded.

The cabin survived the initial explosion long enough for one astronaut to utter the last recorded words, "Uh—Oh!" and long enough for several astronauts to turn on the emergency oxygen supplies to their suits. The cabin plunged for almost three minutes until it hit the ocean, killing the astronauts.

There had been many warning signs that the O-rings were inadequate. Since the solid rocket boosters were recovered after each flight, NASA had had plenty of opportunity to inspect the O-rings, and they had found, on several occasions, that O-rings had nearly burned through. The engineers at Thiokol Corporation, who designed these boosters, had warned NASA not to launch on this cold day, but they were overruled by NASA administrators. The philosophy in the space program had become one of, "Well, if it worked O.K. the last time,

> *Another major item of new equipment needed for the ideal space transportation system during the '70s is the space shuttle—a combination launch vehicle and spacecraft that would operate between the ground and earth orbit. Fully reusable, it would shuttle cargo and passengers to a space station, and return to earth in a runway-type landing.*
>
> *—Wernher von Braun, 1970*

Let us make recommendations to ensure that NASA officials deal in a world of reality in understanding technological weaknesses and imperfections well enough to be actively trying to eliminate them. They must live in reality in comparing the costs and utility of the Shuttle to other methods of entering space. And they must be realistic in making contracts, in estimating costs, and the difficulty of the projects. Only realistic flight schedules should be proposed, schedules that have a reasonable chance of being met. If in this way the government would not support them, then so be it. NASA owes it to the citizens from whom it asks support to be frank, honest, and informative, so that these citizens can make the wisest decisions for the use of their limited resources.

For a successful technology, reality must take precedence over public relations, for nature cannot be fooled.

—Richard Feynman, Nobel laureate and member of the Presidential Commission on the Space Shuttle Challenger Accident

it'll be O.K. this time." Since the Space Shuttle had flown two dozen times successfully, it was assumed that no serious problems remained.

After the accident, a Presidential Commission was set up to investigate the reasons for the accident. It included such notables as Neil Armstrong, the first man on the Moon, Chuck Yeager, the man who had broken the sound barrier, and Nobel Prize-winning physicist, Richard Feynman.

The Commission conducted an extensive, painstaking investigation of the disaster. Members traveled to NASA and industry centers to investigate the manufacture and testing of the Space Shuttle. They studied the data that had been sent back from *Challenger* and the flight recorders recovered from the sea bottom. They grilled NASA administrators and people from industry until they got a consistent story of what happened.

It is fortunate that they were able to pinpoint the trouble precisely. If they had not been able to narrow it down to the O-rings, the entire Space Shuttle might have had to be redesigned at much greater delay and cost.

The Commission discovered that there had been a number of previous problems with these O-rings, which should have warned NASA about the potential danger. When the surviving astronauts learned that on some previous flights, the O-rings had almost burned through and that they had not been informed of the dangers, they were very angry. *Challenger* pilots had not even been told of the Thiokol engineers' recommendation against launching.

The Commission also discovered that often the ground crew personnel were overworked at Cape Canaveral. They found that, for example, on January 6, 1986, the *Columbia* almost launched with too little fuel. An operator who had been working for eleven hours during the third day of 12-hour night shifts accidentally drained 18,000 pounds of liquid oxygen from the Shuttle. It came within 31 seconds of being launched with insufficient fuel. The astronauts would not have been able to go into orbit and would have had to make an emergency landing, possibly in Africa. (This was the mission that carried Congressman Bill Nelson into space. It was delayed a record seven times before launch.)

Precedents

There is a lesson from the *Apollo* program that should help us see our way through those times when it seems that the space program can never recover from the *Challenger* tragedy.

During the early days of the *Apollo* program in 1967, three astronauts were conducting routine tests aboard an *Apollo* module on the ground. A spark ignited the pure oxygen atmosphere of their cabin. Normally, on the ground, ordinary air (one-fifth oxygen, four-fifths nitrogen) was used. In space, the inert nitrogen was omitted and only pure oxygen at one-fifth of an atmosphere pressure was used, the same pressure as we would breathe on Earth if all the nitrogen were taken away. However, for this test, pure oxygen was pumped at one atmosphere of pressure into the cabin—five times normal oxygen pressure.

Many substances that burn slowly in ordinary air burn fiercely in such amounts of oxygen. The interior quickly caught on fire, and because the hatch took too long to open, the astronauts were not able to escape. All three died.

The *Apollo* tragedy set back the NASA program by at least a year and forced the redesign of parts of the *Apollo* capsule. Although this was a traumatic experience to the space program, in the long run the program recovered and went on to land astronauts on the Moon, on schedule. I expect this to happen in the wake of the *Challenger* disaster, which ironically occurred on the day after the anniversary of the *Apollo* fire.

The Soviet record in space is similar, in terms of lives. On one occasion, for example, cosmonauts orbited the Earth successfully and reentered the Earth's atmosphere, seemingly landing safely, but when the ground crew opened up the space capsule, the three cosmonauts were dead. Their hatch hadn't fully closed and the cosmonauts were asphyxiated. On another occasion, according to one report, a huge booster blew up on the launch pad, apparently killing many people. They didn't let this stop them.

After *Challenger*

NASA had gained an image in Congress of a superhuman agency, one that could accomplish anything. The accident proved they were mortal.

For months after the explosion, an atmosphere of gloom pervaded the NASA centers. Several administrators accused of negligence left NASA. Other scientists, engineers, and administrators, seeing their programs delayed by years, also left. Yet, enough people remained who are dedicated to space exploration to ensure that NASA will survive.

After the success of the *Apollo* Moon landings, NASA had become complacent. After all, people said, "If we can put a man on the Moon, we can do anything." NASA relaxed until it became almost an ordinary government bureaucracy. Some members were more concerned with their "perks" and promotions than with the adventure of space exploration. Unfortunately, the rules of Civil Service—under which most NASA centers operate—make it very difficult to get rid of the incompetent or lazy. So with the esprit de corps of a Moon-landing no longer present, there was little to keep NASA operating in the lean and efficient manner that had once been normal.

Then, too, the old German rocket scientists were dying out. Von Braun was dead; the others were reaching retirement age and leaving the agency. Those who took their places were often administrators

*Some historians see history as an accumulation of error. But
history is also the story of the defiance of the unknown and
of what happens when man tries to extend his reach. Such
defiance is necessary because conventional wisdom has
never been good enough to run a civilization. Not all problems are old problems; therefore, new approaches and new
truths have to be discovered. In order to answer the question, "why explore?," then, it becomes necessary to refer to
the phenomenon of human progress. I have a theory that
progress is what is left over after one meets an impossible
problem. The reason it is safer to travel in a Boeing 747
than to sit in your bathtub is that adequate thought has
been given to all the things that can go wrong when you are
in a 747, and not enough thought to what can go wrong in
a bathtub. When you are in a 747, the experts relieve you of
the responsibility for making correct decisions. This is
something that does not happen in your bathtub. What I am
trying to suggest is that the more difficult and complex the
undertaking, the more likely it is that knowledge will be
gained that can be applied more fruitfully far beyond the
undertaking itself.*

—Norman Cousins, author

and others without the practical experience of the old rocket engineers.
Some of these new administrators were more accustomed to flying
a desk than a rocket, more adept at handling paperwork and office
politics than nuts-and-bolts engineering.

The *Challenger* disaster shook up NASA profoundly, and the result
will probably be good in the long run. The agency was reorganized,
and there is now a spirit of determination to make sure the mistakes
of the past are not repeated.

The joints on the solid rocket boosters have been redesigned, and
the safety of the Shuttle and the future Space Station were thoroughly
reexamined by the astronauts whose lives will depend on them. The
improved Shuttles began flying again in 1988.

One thing we must not do is let the *Challenger* incident deter us.
Exploration has always been hazardous. People have always died. The
pilot astronauts have always been test pilots or combat pilots who knew

the risks and accepted them. The development of the ordinary airplane was fraught with crashes and deaths—the price that had to be paid so the rest of us could travel safely from one continent to another.

We must not make the opposite mistake and go overboard in trying to reduce the dangers to zero. This would be impossible, and even the attempt would result in a ludicrously overweight and overdesigned spacecraft. We will never get anywhere if we become fanatics for our safety, just as we would never get anywhere in automobiles if we had waited for them to be perfected to the point where no accidents could occur. We accept the fact that 40,000 people die every year on U.S. roads. American astronauts have traveled millions of miles; and only seven lives have been lost in space. In fact, the most remarkable aspect of the *Challenger* disaster is that it was the first time in the history of the American space program that any human lives were lost in space.

I hope we do not do what many people are suggesting and forbid anyone but the professionals from going into space. We desperately need to send more ordinary citizens into space, people who are not astronauts. We need to send writers and artists, politicians and poets, teachers and TV reporters. The more people who experience the unique vision of this planet that comes from circling it every 90 minutes, the better people will understand our place in the universe and the delicacy of the ball on which we live. Virtually every astronaut has come back changed from the experience, gaining a cosmic perspective that no one stuck on this planet can quite reach.

As long as the Shuttle passengers know the dangers and are willing to accept them, they should go. NASA did have a program slated to send a journalist into space. Newsmen like Walter Cronkite were prac-

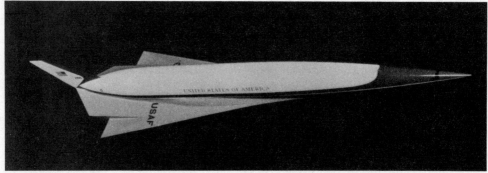

A proposed Boeing design for the American aerospace plane.

tically standing in line for the chance. Since so few travelers in space have been able to share their experiences in vivid language, it is vitally important, I feel, to send journalists out there. Two congressmen, Senator Jake Garn and Congressman Bill Nelson, flew in space on the Shuttle, and this is a good thing. What better way can our leaders get a global perspective than by seeing this fragile world from the perspective of an orbit in which there are almost no boundaries between nations?

The Shuttle disaster has also forced us to reexamine our goals in space. While in the Soviet Union, the space program has tended to move steadily toward expanding the use of outer space, the American program has fluctuated up and down, depending on the public emotions of the moment. This disaster gives us time to pause and think about what we should be doing there.

We have time now to study our options and make decisions. People sometimes ask, "Should we build more Space Shuttles?" The President has made the decision to build at least one more, but the decision to build even more should not be made until we know what we want to do in space.

We must also decide whether to go to the next generation of technology. NASA and the aerospace industry have come up with wonderful designs for an aerospace plane. In a sense, the Space Shuttle is old technology. The materials, design, and the electronics are largely "state-of-the-art" of ten years ago. Much has happened since then. New materials have come along, lighter and stronger. Electronics have shrunk to the point where we can put far more artificial intelligence into the spacecraft than was possible ten years ago. Brilliant new designs have been proposed that would allow an airplane-like vehicle to take off horizontally, accelerating in the atmosphere by using air to burn the fuel, greatly reducing spacecraft weight, since it doesn't need to carry much oxygen. (One-third of the Shuttle's launch weight is liquid oxygen!)

The Shuttle was supposed to be a much more economical way of going into space than the old disposable rockets. But it didn't work out that way because of overoptimistic hopes for its cost and reliability, compromises in the design, and political battles that prolonged its development.

The government (like all other space powers) subsidized launches indirectly, in a way that made it seem that the launches were cheaper to the commercial user than they really were. This is not a good way

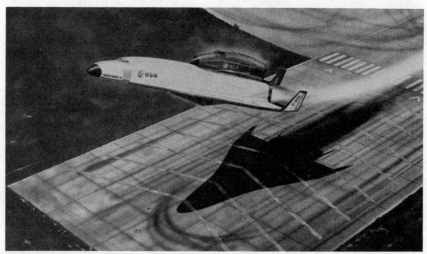

(a)

(b)

Two proposed Western European piloted shuttles:
(a) The *MBB-ERNO Sänger II* two-stage aerospace plane. The first stage would take them to twenty miles altitude: the second would fly on to low-Earth orbit. Both stages would be able to land at any major European airfield.
(b) The European Space Agency's *Hermes*, which would carry pilots on top of conventional rockets.

to put space usage on a commercial basis. Following the *Challenger* disaster, the President announced that, henceforth, commercial space payloads would be largely turned over to private companies using un-piloted rockets. The Shuttle is to be reserved primarily for military and scientific uses.

Formerly, dozens of agencies had to be fought in order to launch rockets privately into space. Some progress has been made during the last few years in reducing such bureaucratic obstacles, but much re-mains to be done. Until the red tape is minimized and the NASA sub-sidies reduced, it will be very difficult for private rocket developers to profitably launch payloads into space. Some in NASA resisted because they saw these companies as competitors for the Space Shuttle, taking away payloads from them. But such competition will, in the long run, be healthy, because competition will keep launch costs to a minimum. The cheaper we make the real cost of space travel, the sooner we will be able to fully develop space for commercial use. This will ultimately make space science cheaper as well.

In the meantime, the Soviets have flown their own space shut-tle, and the western Europeans, inspired by CNES (the French equivalent of NASA), are also designing a small, piloted Shuttle called *Hermes*. The British are studying an aerospace-plane design called *Hotol* (horizontal take-off and landing). Even the Japanese are star-ting to look at similar concepts. Just as the DC-2 airplane was not a great commercial success, but led to the DC-3 that revolutionized air travel for business and pleasure, so too the Shuttle may one day be seen as the spacecraft that paved the way for the first generation of truly practical aerospace vehicles.

And when the names of all the kings and queens and presidents and prime ministers of our time are forgotten, the legacy of the *Challenger* crew will be found in the cities of the Moon, the highways of Mars, and the starships of the galaxy.

THREE

A Nice Place to Visit

EARTH

The average person speaks of Earth with little awareness that it is just one small planet in the vast universe. Most people have never fully realized that it is, in all probability, only one of a multitude of planets as numerous as grains of sand on a beach. This magnitude is hard to grasp when we look at our seemingly endless surroundings, unable to see over the next horizon, rarely traveling to another country—let alone another continent.

Only from space have we begun to look at what our planet really is: a small blue ball, floating in an endless vacuum, moving around an ordinary, insignificant star, just one of hundreds of billions in the Milky Way galaxy. Only from orbit can we see many features too big for us to perceive on the ground. From space, we can see the outlines of huge craters where meteorites once hit this planet, remnants of the forces that first formed our world. Only from space can we begin to get a complete picture of the world's weather and climate. Instead of

If I had to pick one spacecraft, one Space Age development, to save the world, I would pick ERTS [Earth Resources Technology Satellite, renamed LANDSAT] and the satellites which I believe will be evolved from it.

—James C. Fletcher, NASA Administrator

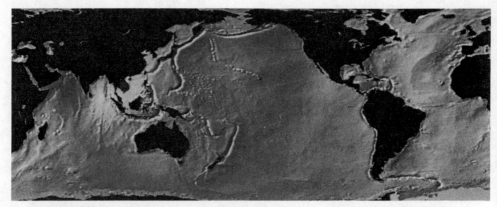

NASA's *SEASAT* pioneered the use of powerful imaging radar to produce the first detailed global picture of the gravitational distortions in the oceans caused by mountain chains and trenches in the depths. Notice especially in the Pacific where the gravitational pull of mountains (some tips of which form islands) piles water higher over them, while the deep trenches produce corresponding dips. We are now beginning to understand how these affect currents that alter our weather and climate.

taking thousands of measurements from random points on the surface of the Earth, we can now take a photograph of an entire hemisphere of the planet in one moment, and watch, heedless of national boundaries, the storms and clouds and rain and hurricanes and snow unfold in mysterious, intricate patterns.

Satellites now routinely allow us to monitor the Earth's crops; we can spot agricultural blights before they spread too far to control. We can find geological resources from space; telltale formations where minerals or oil lie hidden often can be seen from orbit. Scientists have developed increasingly sophisticated spacecraft that can monitor the Earth's surface at many different frequencies, not just with visible light, but with infrared and ultraviolet light, and with microwave frequencies, too. Different rocks and plants, and snow cover, emit distinctive patterns of frequencies that a computer can be programmed to detect. Pollution can also be monitored similarily.

Until we began to watch the Earth from space, we couldn't make many observations essential to our progress that simply cannot be made close up.

People often ask, "Isn't it a waste of money to go into space? Why shouldn't we spend that money here on the Earth instead of throwing it away in space?" But that money *is* being spent on this planet.

It's being invested in helping us find the resources we need to grow as a civilization. It is, in fact, helping us solve many of our global problems.

And that money saves lives. Unknown thousands have been saved by the weather satellites that detect the motion of hurricanes and other storms. Without the advance warnings from these satellites, many people would have died in the terrible storms and hurricanes now routinely observed from space.

One of the most exciting space developments in recent years may even prevent starvation in many countries. Space-borne radar on a Shuttle flight, using a type of radar developed for the old SEASAT oceanographic satellite, was able to penetrate the top layers of sand and vegetation on the Earth to make some remarkable discoveries. This radar revealed, for instance, ancient riverbeds in the Sahara

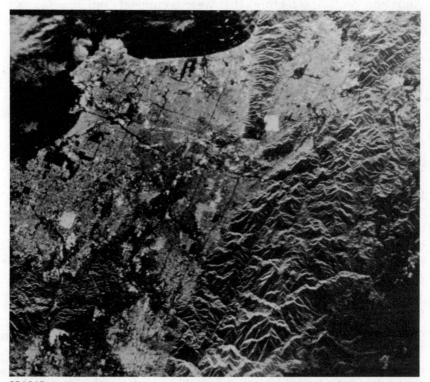

SEASAT radar picture of the Los Angeles basin, showing both artificial and geological structures.

Desert, now completely covered by sand. Though many such water sources long ago dried up, others with greater potential may soon come to light. When the Space Shuttle is flying again, more such observations can be made as it flies over the deserts of the world. We may find hidden water sources with which to help stem the droughts that plague inhabited areas on the fringes of the desert, and thus prevent mass starvation.

Another extraordinary discovery made by this radar was a series of canals in the jungles of Central America, dug by ancient Indians. Overgrown with jungle, they had not been obvious to people on the ground; from space, however, the pattern of canals was easily detected. The more we fly this type of radar, the more we will discover signs of our history, lost cities in the deserts and jungles of the world. NASA has plans to send this radar back into space, and one day we will probably map much of the surface of the Earth this way, discovering ancient ruins and geological features that were completely unknown to us. Indiana Jones will now be an astronaut.

Each passing year sees increasingly sensitive ways to monitor Earth from space, enabling us to discover ever more subtle details on the surface, and below. In the 21st century, people will wonder how it was possible to manage government and industry without data from space.

Planetary Greenhouse

One of the most important things we can do in space is to watch for any changes on our own planet. There is, for instance, a growing suspicion that Earth is getting warmer and may suffer from a "greenhouse effect" like Venus. If so, the climate of the future may be radically different from now, and the effects could start to become apparent in the next decade or two. Droughts may hit places that ordinarily have good rainfall. Shorter growing seasons could cause massive, worldwide food shortages. The Soviet Union, unable to grow enough food for itself even under the best of circumstances, might find its crops substantially reduced.

If this effect continues into the next century, the polar ice caps might melt sufficiently to flood coastal cities. New York and London could become Venices.

The main cause of this alarming trend is the widespread burning of fossil fuels such as coal and oil, and even wood. All these fuels are carbon compounds—organic chemicals—and when they burn, the oxygen in the air combines with the carbon to produce carbon dioxide. Carbon dioxide builds up in a thick layer over the whole Earth, eventually acting much like the walls of a greenhouse, keeping us warmer than we'd otherwise be.

Since the gas is invisible, sunlight passes through it without difficulty and hits the ground, heating up the Earth. Oxygen is invisible, too, but carbon dioxide has an annoying property that oxygen does not; it absorbs infrared radiation. (At room temperature, everything radiates heat waves—lightwaves of infrared light, too red to be seen by the human eye.) The hotter an object is, the more infrared radiation it emits.

In general, Earth maintains its temperature because there is a balance between the sunlight we receive and the heat the Earth radiates back into space. If the Sun were dimmer, the Earth would receive less heat in the form of sunlight, so it would heat up less and radiate a lot less infrared energy into space. If the Sun were brighter, we would heat up more and radiate more infrared into space. We have an equilibrium.

But as soon as you throw a log on the fire, you start to disturb that equilibrium. The carbon dioxide produced by the fire spreads into the atmosphere and traps some of the infrared radiation of the Earth that otherwise would have escaped. Eventually, it will still escape, but that heat does hang around in the atmosphere a little longer because of the carbon dioxide.

So picture this: All over the world, people cook food over fires, burn gasoline in their automobiles, use coal to generate electricity, and put tons of this blanketing gas into the atmosphere from countless other sources. The result, many scientists predict, will be a dangerous greenhouse effect.

And yet this might not be the actual result. Earth is a very complicated planet. More plants will grow, breathing in more carbon dioxide and turning it into oxygen. Pollution puts dust into the air that helps water droplets to condense, forming clouds. Clouds tend to reflect some of the sunlight away from the Earth, and can actually make it cooler.

Other trends will certainly come into play. There are whole jungles being cut down and oceans with marine life being polluted, adding yet more variables to the complex interaction between the atmosphere and the planet's surface.

> *Despite our highly advanced civilization, we still cannot predict the climate in any dependable way. Progress is being made, however, and one day man may be able to foresee potentially catastrophic changes in his biosystem. With this knowledge he may be able to avert such changes by adapting to them gradually or in some cases by preventing them.*
>
> *The main obstacle to gaining control over man's long-term climatic destiny is the sheer complexity of the total climatic system of the earth. Global climate consists of five subsystems [atmosphere (air), hydrosphere (water), cryosphere (ice), lithosphere (rock) and biosphere (life)] that interact through a great variety of processes.*
>
> *—International Institute for Applied Systems Analysis, Austria*

Recently, a "hole" has opened up above Antarctica in the ozone layer that protects us from deadly solar ultraviolet. Freon gases from refrigerators, air conditioners, and spray cans may be the culprit. We need to monitor this development to see whether it is a threat to our ecology.

The whole picture is so complicated that no one can confidently predict what's going to happen. On top of all the factors immediately visible, we know there have been ice ages and tropical ages in the distant past, long before humanity began to fiddle with the planet. Clearly we can expect changes in Earth's climate, independent of our actions.

The only way we can know whether we are going to trigger climatic disaster or not is to study the atmosphere and the oceans, the forests and farmlands, from space. An entire army of meteorologists out there with their thermometers and wind gauges can do but a tiny fraction of what one good weather satellite can.

Before this era, we couldn't see the forest for the trees. Now we can not only see it, we can map it, take its temperature and check it for Dutch elm disease. Soon, we'll be able to give the entire Earth a periodic physical check-up, and perhaps keep our planet from becoming very sick.

FOUR

Star Wars

THE STRATEGIC DEFENSE INITIATIVE

In 1983, President Reagan announced a startling proposal, as controversial and seemingly science-fictional as President Kennedy's plan to put a man on the Moon. This was Reagan's "Star Wars" speech, in which he proposed building a shield to protect the U.S. against missile attack.

The concept had come about largely because of an idea that Edward Teller, widely known as the Father of the H-bomb, presented to Reagan. He proposed building a kind of "death-ray" that could destroy missiles in space. It would use a nuclear explosion to produce intense X-rays that could zap Russian missiles heading toward the U.S. Since this was about five years after the enormous popularity of the film *Star Wars*, the nickname was inevitable. Its detractors regarded the term as an insult, while some of its supporters felt that the name should be worn

> *When a distinguished but elderly scientist states that something is possible, he is almost certainly right. When he states that something is impossible, he is very probably wrong.*
>
> —*Arthur C. Clarke, author*

There has been a great deal said about a 3,000 miles high-angle rocket. In my opinion such a thing is impossible for many years. The people who have been writing these things that annoy me, have been talking about a 3,000 miles high-angle rocket shot from one continent to another, carrying an atomic bomb and so directed as to be a precise weapon which would land exactly on a certain target, such as a city.

I say, technically, I don't think anyone in the world knows how to do such a thing, and I feel confident that it will not be done for a very long period of time to come. . . . I think we can leave that out of our thinking. I wish the American public would leave that out of their thinking.

—Vannevar Bush, Director,
Office of Scientific Research and Development, 1945

proudly, as it symbolized the fact that so many of the inventions of science fiction had already become reality—robots, lasers, and spaceships, to name just a few. Officially, it was known as the Strategic Defense Initiative (SDI). Here, we'll call it Star Wars without prejudice one way or the other, and will look at its pros and cons.

For decades the U.S. and the Soviet Union have relied on one principal means of preventing war—the threat of complete annihilation of the opposition. For much of history, major wars have occurred at intervals of 20 to 30 years. Indeed, it was a scant two decades between the end of the first World War and the start of the second. But despite frequent hostilities between the West and the Eastern bloc, more than 40 years have passed since the end of World War II without another global conflict.

Probably the only reason we have not had such a war is that even the most fanatic aggressor cannot tolerate the thought of being a victim of the opponent's modern weapons, each of which could contain the power of millions of tons of TNT. (Compare that with the most powerful non-nuclear bomb ever dropped in war, the British Grand Slam of World War II—it weighed a mere eleven tons.)

What changed everything started way back in 1905 when Einstein

published his most famous equation, $E = mc^2$, the energy source of
the atomic bomb. The U.S., with help from its allies, developed the
bomb and ended the war with the destruction of Hiroshima and
Nagasaki.

In 1946, while the memories of Hiroshima and Nagasaki were fresh
in the world's mind, the U.S. drew up a policy giving the United Nations absolute authority over all nuclear materials that could be used
to make weapons. The Soviet Union, however, under Stalin, rejected
this concept, and the world lost its greatest chance to defuse the Bomb.

The weapons race had begun. The Russians had no intention of being left behind and, to the astonishment of most, produced their first
atomic bomb only four years after the war. Soon England, France,
China, and India joined the club. The bomb became compact enough
to be carried aboard not only airplanes but on the descendants of the
German V-2 rockets. Today there are about 50,000 nuclear weapons
in the world, with an energy total of about a million times that of the
Hiroshima bomb.

The race to build bigger and more effective weapons culminated
in the most powerful bomb ever detonated, equivalent to 57 million

*. . . by converting hydrogen bombs into hitherto
unprecedented forms and then directing these in highly
effective fashions against enemy targets would end
the MAD [mutual assured destruction] era and commence
a period of assured survival on terms favorable to the
Western Alliance.*

*—Edward Teller, "the father of the H-bomb,"
private letter to President Reagan, 1983*

*American political and military leaders should publicly
acknowledge that there is no realistic prospect for a successful population defense, certainly for many decades, and
probably never.*

—Harold Brown, Secretary of Defense to President Carter

Since the late 1960s, the Soviets have been involved in research to explore the feasibility of space-based weapons that would use particle beams. We estimate that they may be able to test a prototype particle beam weapon intended to disrupt the electronics of satellites in the 1990s. A weapon designed to destroy satellites could follow later. A weapon capable of physically destroying missile boosters or warheads probably would require several additional years of research and development.

It is still uncertain whether ground-based charged particle-beam weapons are feasible—that is, whether the beam will propagate in the atmosphere. A space-based neutral particle-beam weapon, however, would not be affected by the atmosphere or by the earth's magnetic field.

Soviet efforts in particle beams, and particularly on ion sources and radio frequency quadrupole accelerators for particle beams, are very impressive. In fact, much of the U.S. understanding as to how particle beams could be made into practical defensive weapons is based on Soviet work conducted in the late 1960s and early 1970s.

—Soviet Strategic Defense Programs, *U.S. Department of Defense*

We will have "Star Wars" or arms control. We can't have both.

—*Clark Clifford, Secretary of Defense to President Johnson*

tons of TNT, exploded in 1963 by the Soviets. In the same year, Soviet Premier Nikita Khrushchev announced that they had a 100-megaton bomb.

Technology got us into this mess, and it's natural to ask whether technology can also get us out of it. Many have tried to find scientific solutions during the past few decades. In the 1960's, an ABM (Anti-Ballistic Missile) defense system was proposed that would have tried to shoot down incoming missiles with other missiles. After limited systems were built in the Soviet Union and the U.S., the ABM treaty was signed, stopping the construction of further such systems, although not preventing research on them.

While the U.S. was preoccupied with the Vietnam War, the Soviets developed anti-satellite weapons. Western radar operators watched on screens as the Russians put up satellites that maneuvered close to other Soviet satellites and blew them up. Although these were not missiles, they were the first hints of real Star Wars: spacecraft shooting down spacecraft.

Then, in 1977, came a controversial interpretation of spy-satellite pictures of a mysterious installation in the Soviet Union. Major General George J. Keegan, the retired former director of USAF Intelligence, announced that it was virtually a Buck Rogers death ray: a particle-beam weapon. If he was right, the Russians had built an experimental device on the ground that could accelerate electrically charged

Artist's conception of the Soviet Tyuratam Space Complex, showing antisatellite interceptors. The inset shows pellets being deployed to destroy a satellite.

> . . . we can place into space the means to defend these
> peaceful endeavors from interference or attack by hostile
> powers. We can deploy in space a purely defensive system
> of satellites using non-nuclear weapons which will deny any
> hostile power a rational option for attacking our space
> vehicles or for delivering an effective first strike with
> ballistic missiles. Such a global ballistic missile defense
> system is well within our present technological capabilities
> and can be deployed in space, in this decade, at a cost less
> than other available options.
>
> —Lt. Gen. Daniel O. Graham (USA, Ret.), Project Director,
> High Frontier
>
>
> Our opposition to entrusting the survival of the planet to a
> Death Star is not because it is an alternative, but because it
> is the wrong alternative. Their Maginot Line in the sky can-
> not provide Mutual Assured Survival. It cannot even pro-
> vide Assured Survival for Fortress America.
>
> —Robert M. Bowman, President,
> Institute for Space and Security Studies

atoms to nearly the speed of light, zapping any missile or satellite that came within its beam. Many disagreed with his interpretation, both inside and outside of the defense establishment, but his claims served as a catalyst to the American ideas that became Star Wars.

And so now, as the world lives on the edge of Armageddon, countless scientists, politicians, and ordinary people have tried to devise a way out of this terrible dilemma that seems to point toward possible annihilation of civilization. Their attempts basically divide into two paths: the political and the technological. The political solutions would try to defuse the situation by nuclear disarmament, or at least a reduction in the number of nuclear weapons in the world. The technological path is Star Wars. Both paths are covered with potholes and are dangerously close to cliffs, because if not followed with great care and foresight, either approach could lead to World War III.

The political approach sometimes overlooks the likelihood that, without the threat of global annihilation, the small confrontations that flare up chronically in the Middle East, Africa, Central America, and Asia might escalate into another World War.

But the technological approach has similar risks. One of them is that no shield in the foreseeable future would be perfect. Even if the shield were 90 percent effective, 10 percent of the Soviet arsenal is enough to obliterate civilization in North America. The Soviets could simply make even more weapons to overwhelm defenses, adding to the already horrendous stockpile of weapons in the world.

The critics of Star Wars say that we're never going to solve these problems with more technology; all we'll do is create more problems. The advocates reply that we can't trust the Russians, so we might as well concentrate on defending ourselves.

To what extent are these technological ideas fantasy? Can we actually build an umbrella against incoming missiles? The whole debate about Star Wars has been filled with hysteria, grandiose claims, and misinformation—on both sides.

How do you defend against a missile? In theory, it's surprisingly easy. A cannonball could destroy a hydrogen bomb.

An H-bomb is carried in an Inter-Continental Ballistic Missile (ICBM) at a speed on the order of 10,000 miles per hour. If a cannonball were shot by a similar ICBM into the missile, the relative speed would be

---

Research based on the new technologies of missile defense—the ultra-compact computer, the laser and other sophisticated devices—offers the promise of an end of the nuclear nightmare.

—Robert Jastrow, founder, NASA Institute for Space Studies

The Soviets could penetrate whatever defense was technically feasible if they chose . . . to do so.

—Robert S. McNamara, Secretary of Defense to Presidents Kennedy and Johnson

thousands of miles per hour, and a metal ball hitting that fast would almost certainly destroy the missile completely.

But in practice, it's immensely more difficult. Space is so vast, even a hundred miles above the Earth's surface, that it would require an enormous network of detectors and rockets to do the job, and they would have to be extremely precise. You're aiming at a target that's moving at you in a complicated path thousands of miles away. This "cannonball" type of defense is known as a kinetic-energy weapon, because it simply uses the brute force of the energy of the anti-missile to destroy the target. It's become the preferred near-term concept, though sometimes ignored by the media, perhaps because it isn't as

> By taking an optimistic view of newly emerging technologies, we concluded that a robust BMD (ballistic missile defense) system can be made to work eventually. The ultimate effectiveness, complexity, and degree of technical risk in this system will depend not only on technology itself, but also on the extent to which the Soviet Union either agrees to mutual defense arrangements and offense limitations, or embarks on new strategic directions in response to our initiative. The outcome of this initiation of an evolutionary shift in our strategic direction will hinge on as yet unresolved policy as well as technical issues.
>
> —James C. Fletcher, Chairman, Defensive Technologies Studies Team, advisory panel to President Reagan, 1984

> In the distant future, [Star Wars] will be a technical possibility. But it will always be impossible from the military-strategic point of view, since any strong opponent with a sufficiently high level of technology can always overcome the technical achievements of the other side at all stages and he won't even have to invest as much or as many resources as are being invested as the creator of SDI.
>
> —Andrei Sakharov, "the father of the Russian H-bomb" and Soviet dissident

. . . the problem of integrating human behavior and political institutions with rapidly evolving technology may prove to be the most difficult and crucial challenge to humanity's long-term survival. The teams' failure to resolve fully all the issues raised in this paper should come as no surprise: they will have to be debated with knowledge and impartiality for years to come. Indeed, one of the main values of the president's Strategic Defense Initiative may have been to call attention to the dilemma facing us, and to inspire us to make use of our technical talents in innovative ways. We must not make decisions on the basis of preconceived notions or, still worse, prejudice, but must do so on the basis of knowledge of the facts and an understanding of their broad implications. The purpose of the Strategic Defense Initiative is to provide the hard data as well as the in-depth insight to allow truly informed decisions.

—Gerold Yonas, Chief Scientist, Strategic Defense Initiative Organization, U.S. Department of Defense

The natural response of either of the superpower competitors to the deployment of defenses will be to expand offensive forces.

—James Schlesinger, Secretary of Defense to Presidents Nixon and Ford

glamorous as the more science-fictional techniques. Also, there are many difficulties in getting the weapon to the target in the few minutes between ICBM launch and hit.

Among the more high-tech approaches are techniques involving energy beams— "death rays." There are small working models of these in the laboratory right now. One is the laser beam, and another is the particle beam. Lasers are just lightwaves amplified into large energies, forming extremely sharp beams. Any kind of electromagnetic energy could, in theory, be used to make a laser: infrared, ultraviolet, or even X-rays, not just visible light. The reason lasers are important in this game is that a laser beam is so tightly beamed that it doesn't spread out much over large distances. In theory, it would be possible to aim a laser at a missile hundreds of miles away and hit it. It has the virtue, too, that light travels at thousands of times the speed of a missile, so can outrun it readily. But there are severe practical problems with finding the target and focusing that energy over long distances. For one thing, Earth's atmosphere tends to "defocus" a laser beam.

Particle beams resemble lasers in their use, but require different techniques to operate. Most of us have small particle beams in our homes—they're called TV sets. In each television set there is a picture tube. The far end of the tube has a little "gun" that boils off electrons by using a hot metal filament very much like that inside a light bulb. High voltage, generated by the set, accelerates those electrons until they go flying out of the electron gun and smash into the front of the tube, which is painted with chemicals that glow when hit.

Similar techniques can generate much more powerful beams of particles that would travel through the atmosphere and avoid some of the problems of lasers, although it's much harder to keep the beams from spreading and hence losing their destructive power.

Either weapon is incredibly hungry for energy. If you keep such a weapon on the ground, you can supply it with any convenient power plant, such as Hoover Dam. But then you can't see the missiles until they are very close—unless you use mirrors in space (a real possibility). Or you could put the weapon in space, but then it's much harder to supply the energy. There would probably have to be enormous structures up there to generate the power, and the orbiting weapons would themselves be highly vulnerable to attack.

Orbiting weapons also face a new kind of vulnerability. An atomic weapon going off in space can zap many of the satellites in orbit around the Earth, even thousands of miles away. There's no Earth to absorb the radiation, and on top of that, a nuclear explosion generates a great pulse of electromagnetism that can blow out satellite electronics. This is one area in which the U.S. has become more vulnerable than the Soviet Union. Because we love high tech—and we're good at it—we rely more on ultrasophisticated electronics in space for communications and intelligence than do the Russians. However, there are ways to harden and shield the electronics so they're less sensitive to radiation, and research in this area is proceeding rapidly.

How do you shoot down a Russian rocket? Star Wars breaks the prob-

In previous wars, when the deterrent power of armies, navies and, more recently, air forces has failed, the resulting conflict has usually been wretchedly damaging and destructive; but it has not been irretrievably annihilating. A nuclear war would be. It is for this reason that the posture of Mutual Assured Destruction is, and always has been, a dangerous and unsatisfactory concept. It has, it is true, prevented a war between the two great conflicting political and economic systems for forty years; but the potential cost of a breakdown is so crushing that it would be irresponsibly foolish not to seek constantly for some more reliable means of ensuring peace with security. As The Times declares, ". . . it must be right to prefer a defensive system, albeit an imperfect one, than to continue with the arid increase of mutual assured destruction."

—Lord Alun Chalfont, Chairman, House of Lords
All Party House Defense Group

The Star Wars policy is ill-advised on both technical and strategic grounds. There is virtually no chance that an invincible shield envisioned by President Reagan can be developed. Yet the pursuit of this appealing mirage will, ironically, make us less rather than more secure: It will escalate the arms race, reduce stability, and feed a new cycle of mutual suspicion and fear between the superpowers.

—The Union of Concerned Scientists

lem down into different stages of the missile's path. The easiest time to zap a Soviet missile is when it first launches—the boost phase. This is when it moves slowly. It's like shooting a duck in a barrel. It's also the most dangerous time to start shooting, because if the "missile" turns out to be an ordinary rocket and not an attack, Star Wars could trigger the very war it's supposed to prevent.

After the missile enters space, it's moving much faster and is harder

As an astronaut, I have long deplored our failure to make maximum use of our space technology advantages. The High Frontier ["Star Wars"] concepts pull both military and non-military uses of space into one coherent U.S. strategy.

—Edwin "Buzz" Aldrin, Apollo 11 *astronaut*

Given that the SDI program is still in its early stages, it is difficult to estimate the final costs of deploying and maintaining a multilayered durable and effective BMD system. Tentative estimates that have been made range from $100 billion to more than $1000 billion (one trillion), but these must remain speculative until more is known of the configuration of the BMD system.

What can be said with some certainty is that deploying and operating strategic defenses against ballistic missiles, involving dozens if not hundreds of complex weapons platforms and sensing systems in space, extensive command-and-control networks, as well as numerous space vehicles to deploy and repair the system, will be a very expensive proposition. If comprehensive defenses against manned bombers and cruise missiles are deployed as well—an effort former Secretary of Defense James Schlesinger has estimated could cost $50 billion annually—then the overall cost could well be prohibitive. . . .

In sum, despite remarkable advances since the 1960s in BMD related technologies, there are major uncertainties surrounding the ultimate feasibility of deploying and maintaining strategic defenses against ballistic missiles. Even if many of the remaining technical obstacles are overcome in the areas of weapons systems, optics, sensors, and guidance systems, there will still remain the issues of countermeasures, survivability, cost-effectivness, and the integration of strategic defenses with overall U.S. military capabilities.

—Hans A. Bethe, Nobel laureate, Cornell University; Jeffrey Boutwell, Committee for International Securities Studies, American Academy of Arts and Sciences; and Richard L. Garwin, IBM Fellow, Thomas J. Watson Research Center to detect. Then it may release a dozen warheads that can travel to different targets. Decoys are also mixed up with them, so the big headache to Star Wars is to try to figure out which objects are decoys and which ones are the real McCoy. If there were enough energy in the system, perhaps the decoys and the fakes could all be destroyed indiscriminately, but that objective would multiply greatly the already major difficulty of supplying necessary power. And, of course, it may be much cheaper for the Russians to make decoys than it is for us to make systems capable of shooting them all down.

Another problem with Star Wars is that it does nothing to protect us against cruise-missiles that travel just above the ground like airplanes, nor does it help against bombs smuggled in on boats or trucks, and nuclear weapons are now so compact that some can be carried in a backpack. Even if Star Wars works, it's not the answer to all our defense problems.

Spy in the Sky

The ultimate problem of civilization—preventing the final world war—may be solved by monitoring Earth from space. MIT Professor Jerome Wiesner, former science adviser to President Kennedy, has recently revived and updated an old proposal for an international agency to play watchdog for the human race.

A PROPOSAL FOR AN INTERNATIONAL ARMS VERIFICATION AND STUDY CENTER

This paper proposes that interested nations collaborate to create an International Arms Verification and Study Center, through which they could help bring under control the nuclear arms race and the threat it poses to all life on the planet. The Center would do this by making arms limitation steps more acceptable through verification measures that would reduce the likelihood that violations of agreements would be undetected. . . .

The most effective non-invasive means for monitoring the deployment of weapon systems involves the use of high-

resolution visual reconnaissance provided by satellite borne cameras. The existence of fixed installations of ballistic missiles and production facilities for most weapon systems can be detected with a high degree of confidence by existing systems; the same thing is true for submarines carrying ballistic missiles. At the present time it is unclear to what degree deployment of mobile missiles can be accurately monitored. Monitoring cruise missiles on land or sea is even more difficult. This is a technical challenge presently receiving considerable attention, but in the end adequate control of these weapons may require actual on-site inspection. Here a mixed, international team of inspectors may be more acceptable than the nationals of the opposing superpowers. . . .

Unlike the situation in 1978, when the French proposal for an international reconnaissance satellite system was put forward and then frustrated by superpower reluctance, it is now within the capability of the midpowers to create one with their own resources. . . .

The superpowers together spent several hundred billion dollars on their military systems. It is estimated that together the nations of the world will spend almost one trillion dollars in 1986. The shared cost of the proposed new Center would be a small fraction of these amounts combined, and would provide much more leverage in the prevention of the most serious threat that faces the citizens of these nations. It would be a major step towards world sanity. . . .

The Sword of Damocles hangs on two unpredictable lengths of thread, the rationality of leaders, and the reliability of machines.

The only way to eliminate this unprecedented threat to life on the planet is to drastically reduce the number of nuclear weapons. All peoples in all nations are at risk and they must all help to eliminate this threat. They can help move the world toward survival through common security by backing the establishment of an International Verification System as a starting point for a common security system.

—Jerome Wiesner,
Science Adviser to President Kennedy, MIT, 1986

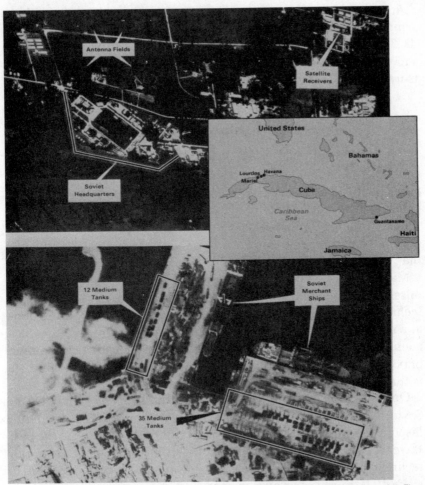

An example of the detail visible by modern reconnaissance photography. The top shows a Soviet intelligence-collection facility at Lourdes, Cuba, the largest outside the USSR. The bottom is the Mariel port facility.

His proposal would create an International Arms Verification and Study Center that would operate its own spy satellites and ground stations, independent of the U.S. and the Soviet Union. For the first time, this would enable the whole world to learn without a political bias whether or not a nation was cheating on arms-limitation treaties. Not only might such an agency be used to reduce tensions between small nations, but it could give humanity the first objective means to cut through superpower lies and propaganda.

At present, both the U.S. and U.S.S.R. often claim the other violates arms treaties, and each dismisses the other's accusations as propaganda. This Center would be much harder to ignore. Independent of the superpowers, it would be a neutral observer to help make arms treaties truly work.

Star Wars or Star Peace?

The Star Wars approach presents problems of enormous complexity and difficulty, but if it saved us from World War III, it would be worth every penny. On the other hand, if it only leads us into another expensive arms race, we will have wasted vast resources on nothing.

In my opinion, both sides of the Star Wars debate have become so convinced of their righteousness that they have lost sight of the fact that we simply don't *know* what the answers are.

Maybe it is possible to build an effective defense against incoming missiles. Probably in the 21st century, the natural evolution of technology will make this possible. The question is: Can it be done at a reasonable cost in a reasonable time without giving either side a temporary military advantage that could upset the delicate balance of power that's kept us from World War III? The only way to find out is through more research.

On the other hand, there have never been so many signs that the Soviet Union is truly willing to make real compromises, including inspection on the ground, which it has strenuously opposed in the past. If we were to ignore the diplomatic possibilities in favor of the purely technological, we might be throwing away our greatest opportunity since 1946 to save the human race.

> *Anyone who thinks the Soviet Union is lacking a comparable SDI program should think again. . . . Since the early 1970s the USSR has had the world's only anti-satellite system. . . . A particle beam research program comparable with that of the United States is also under way.*
>
> —*Bernard Blake, editor of* Jane's Weapons Systems

According to some experts, it was the Star Wars threat that brought the Soviets to the bargaining table in 1986. From the Russian point of view, a space-based defense would let the U.S. take advantage of its enormous high-tech prowess, putting the Soviet Union at a severe disadvantage. This makes Star Wars the greatest bargaining chip we've ever had.

Diplomacy, accompanied by reasonable doses of hardheaded skepticism and caution, seems to be bringing a major breakthrough in this arms race that threatens to bankrupt our whole civilization—if it doesn't destroy us first. The amazing sight of Soviets watching Americans dismantle nuclear weapons, and vice versa, gives new hope to the possibility of drastically reducing the danger of World War III. A result of the Intermediate-Range Nuclear Forces (INF) Treaty between the U.S. and the Soviet Union, it is a breakthrough of the first magnitude. If we can continue to engage in mutual, verifiable reduction of weaponry and the withdrawal of soldiers, then even if Star Wars is as good as its advocates claim, we may not need it, except perhaps for a small system for limited defense against accidental launches.

At least for now, it looks as if the combination of technology and diplomacy may reduce the threat of war for the near future. That seems to be the best we can hope for right now, given the unending history of warfare on this poor, battered planet.

And the Wiesner spy-in-the-sky proposal could be just the breakthrough we need to get us moving toward more sweeping, long-term reduction in our planet's ability to self-destruct.

FIVE

Captain Kirk, Here We Come!

HUMANS vs. ROBOTS IN SPACE

To an astronomer, a robot probe landing on another world, sending back pictures and data, is one of the most exciting possible events. But it's difficult for the average person to relate to a robot, or to be excited by one. When humans landed on the Moon, practically everyone was glued to a television set. When earlier automatic *Surveyor* spacecraft landed there, the only breathless witnesses were science-fiction fans and aficionados of space exploration. As long as space exploration is paid for by the taxpayer, astronauts must and will have an important role.

But there is more to human space exploration than simply putting someone "up there" with whom we can identify. Humans are more sophisticated computers than any yet developed. They can handle circumstances impossible to anticipate. When something goes wrong, you don't want a computer that's been programmed to respond in only limited ways. Humans can spot things going awry, they can fix them, and they can make discoveries of unanticipated phenomena.

The Russians, who know this very well, hold virtually all the world records on the amount of time spent by humans in space. Altogether, they have accumulated about twelve cosmonaut-*years* of time in

> *The success of the Space Shuttle and the advent of the Space Station focus our attention upon a basic question: Why do we have a civil space program? Why do we concern ourselves with rockets and satellites? My own view is that we have a space program not because of the excitement the program engenders and certainly not because the Russians have one. You have a space program because there are things you can do in space that you can't do on the ground. What things? Astronomy, materials research, navigation and communications, Earth observations, are all activities that lend themselves to space and there are others. Space has become simply a place where we do useful things.*
>
> —*Rep. Bob Traxler, U.S. House of Representives*

space—two and a half times as much as the U.S. The world-record-holder for time in space, cosmonaut Leonid Kizim, has spent a total of more than one year in orbit.

The Robotic Columbus

The great explorers of history—Columbus, Magellan, Captain Cook, and their compatriots—would not have gotten far had they been robots. Had Columbus been programmed like a computer, he would have stopped and returned a failure after his calculations indicated he should have seen land but did not. He would not have discovered America.

Darwin did not set out to discover evolution when he journeyed forth on the *Beagle.* His intention was simply to observe the animal and plant life of new lands. He could never have guessed—nor could a committee of scientific experts—that in the process of cataloging all the plants, animals, and fossil rocks he found, he would put these clues together into a theory that would revolutionize biology.

Of course, today's astronauts are a different breed from such independent leaders of small crews as Columbus, or a lone genius like Darwin, who traveled for years without making any contact with the home base. Today's space explorers are just the tip of a pyramid made up of thousands of technicians, scientists, computer programmers, managers, and secretaries. And the astronauts are in constant contact with that network.

Columbus did not have to compete with robots. Astronauts do. Robotic spacecraft may not be as smart as humans, but they don't require expensive, complex life-support systems that greatly add to the weight of their craft. And we can tolerate a lower level of technical reliability when there's no risk to human life,which cuts down on the stringent requirements for safety that also add to the complexity and weight—and cost. Complex requirements are the reason the Space Shuttle is as big as it is, and why it often cannot launch on schedule.

Clearly, there are some functions that are best left to machines. Those spacecraft that don't require much artificial intelligence to operate, like weather and spy satellites, can often be made at lower cost than piloted spacecraft. But even these break down, and if they can be rescued by a shuttlecraft, they can repay the expense of a piloted vehicle.

Astronauts servicing the Hubble Space Telescope, soon to be launched.

The question is, can you invent a machine that will walk? Of course you can! I've seen machines that can walk, but they're usually merely laboratory demonstrations. These machines might have very specialized uses, but I don't think they can ever really take the place of walking. We walk so easily that it makes no sense to kill ourselves working up a machine that will walk. And, as far as computers and human beings are concerned, it is wasteful to develop a computer that can display a human variety of intelligence. We can take an ordinary human being and train him, from childhood on, to have a terrific memory, to remember numbers and partial products, and to work out all kinds of shortcuts in handling addition, subtraction, multiplication, division, square roots, and so on. In fact, people have been born with the ability; they are mathematical wizards who can do this sort of thing from an early age, and sometimes they can't do anything else. But once you train that ordinary human being, what do you have? You have a human being, which you've created, so to speak, at enormous effort and expense, who can do what any cheap two-dollar computer can do. Why both? In the same way, why go through the trouble of building an enormously complex computer, with complicated programming, so that it can create and write a story when you have any number of unemployed writers who can do it and who were manufactured at zero cost to society in general, by the usual process. To sum it up, I think we can be certain that no matter how clever or artificially intelligent computers get, and no matter how much they help us advance, they will always be strictly computers and we will always be strictly humans. That's the best way, and we humans will get along fine.

The time will come when we will think back on a world without computers and shiver over the loneliness of humanity in those days. How was it possible for human beings to get along without their friends? You will be glad to put your arm around the computer and say, "Hello, friend," because it will be helping you do a great many things you couldn't do without it. It will make possible, I am sure, the true utilization of space for humanity. When we finally do extend the living range of humanity throughout near space, possibly throughout the entire solar system and out to the

> *stars, it will be done in tandem with advanced computers that will be as intelligent as we are, but never identically intelligent to humans. They will need us as much as we need them. There will be two, not one of us. I like that thought.*
>
> *—Isaac Asimov, author*

But astronauts have to compete with increasingly intelligent machines. Computers in the modern sense have existed since World War II; each year they have become more powerful, cheap, and compact. A whole new field with the marvelous science-fiction name "artificial intelligence" (AI) has appeared along with these developments. Scientists involved with AI teach computers to recognize objects, to make decisions based on imperfect data, and to simulate the reasoning of experts in such diverse areas as medical diagnosis and oil exploration. Computers are now learning how to "think," and this advance promises to revolutionize spacecraft design. Robots have been experimentally developed that could go to Mars, travel around the countryside, avoid falling off cliffs or banging into boulders. But they're still a long way from having the intelligence of even a child.

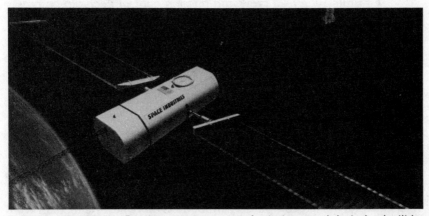

The Industrial Space Facility, a space-manufacturing module to be built by Space Industries, Inc. and Westinghouse Corp. Privately owned, it will be used for the research and manufacture of materials, and will be serviced by astronauts.

There are many ways in which spaceflight conditions are likely to affect dynamics within small groups. Historically, members of isolated and confined groups have had difficulty containing interpersonal conflict without resorting to social withdrawal. Direct training in interpersonal relations may be of some help in this regard. In isolated and confined groups, conformity and compliance pressures are likely to be high; this can make it difficult for individuals to offer creative solutions to problems. Conformity pressures can result in the rejection of the nonconformist, a result which may be unacceptable under spaceflight conditions. Established groups would seem to offer some advantages for long-duration missions. Ideally, a crew entering space would be sufficiently "old" that its members will have achieved a high degree of interpersonal coordination, but sufficiently "young" that they will not have become bored with one another. Space stations and other space environments that are likely to have crews of rotating membership will have to grapple with the problem of assimilating new members into the crew.

Many isolation studies have revealed overt hostility between crewmembers and external monitors or authorities. Although there is general agreement that conflict can have a functional aspect, damaging conflict between organizational units such as a crew and mission control needs to be prevented or at least contained. Persons who serve as interfaces between the crew and other groups or organizations play a critical role in determining the presence or course of intergroup hostility.

—Mary M. Connors, Albert A. Harrison, and Faren R. Akens,
Living Aloft, *NASA Special Publication 483*

Many people have an exaggerated idea of what computers can do today. They've seen too many almost-human robots in science-fiction films. Actually, what computers *cannot* do is more surprising than what they *can*. When they were first invented, scientists expected that it would be easy to make computers do what children do and difficult to make them do what scientists do. But, strangely, the opposite was true. It was relatively easy to teach computers to solve problems in calculus. Yet when they attached twin TV cameras to the computer to simulate the three-dimensional vision our two eyes use, they found it was very difficult to make a computer recognize objects or even play with building blocks; to see, for instance, that block A is a sphere and B is a cube, and to know that the sphere can be set on the cube but not vice versa.

The problem is that the real world is very complicated. Even a child's mind is a much more powerful computer than any supercomputer yet built. The mind can analyze in a split second an enormous amount of data received by the eyes, and it can react to unexpected events. For instance, returning to the above scenario, a computer programmed to handle only two blocks would go crazy trying to deal with a third. And if we have this much trouble in a cozy laboratory, imagine how difficult it is to make a machine respond to the alien conditions of outer space.

There appears to be a number of important psychophysiological variables that relate to an individual's ability to adapt to spaceflight. For example, space sickness or space adaptation syndrome affects about one-half of all space travelers, primarily during the early days of a mission. Earth-based studies, although far from conclusive, suggest that . . . age, gender, and personality traits, could be predictive of an individual's susceptibility to space sickness. Sensory conflict theory continues to be helpful in directing research efforts in the general understanding of why space sickness occurs and how to simulate it on Earth. . . .

. . . one intriguing line of evidence suggests that the highly athletic individual may present no advantage in withstanding the effects of weightlessness and may even be at a disadvantage compared with an average, healthy individual. Similarly, there is evidence to suggest that older individuals may handle certain types of physiological stress better than younger ones. These and related findings could have important selection implications.

—Mary M. Connors, Albert A. Harrison, and Faren R. Akens,
Living Aloft, *NASA Special Publication 483*

There is no simple answer right now to the issue of astronauts vs. machines, and those scientists who attack the piloted space program are sometimes as blind as those who think we need people up there to do everything. The human presence may add cost to a mission, but it can reduce costs in the long run. For example, the Hubble Space Telescope that will be launched when the Shuttle resumes flying is designed to be repaired and refurbished by astronauts. A billion-dollar spacecraft could become scrap metal if it weren't for people in space.

Rescue Mission

The usefulness of astronauts in space has been shown many times. One of the most dramatic was in 1973, when our first, limited Space Station was launched—*Skylab. Skylab* was to be the first place in which American astronauts could live for extended periods of time. Following the triumphs of the *Apollo* program, three sets of astronauts were to be launched over a period of a year, each group to live in space for up to three months.

> . . . there is certainly clear biomedical evidence that deconditioning [from extended time in zero gravity] can be dangerous to the safety and survival of the astronaut, if measures are not taken to limit or reverse the deconditioning process. The successful completion of two 6-month-duration missions within an 18-month period by a Soviet cosmonaut . . . does allay many of the fears for future missions. However, the capacity of different groups of humans to resist pressures will continue to be an area of obvious concern.
>
> Traditionally, our approach to reducing physiological deconditioning in space has been to select men in top physical condition and maintain this condition through exercise. Rigorous preflight and in-flight conditioning programs have been maintained under the assumption that the better the astronaut's physical condition, the greater is his overall resistance to the stresses of spaceflight. In-flight exercise does appear valuable in reducing muscle deconditioning. The comprehensive exercise program used during Skylab missions was effective in preventing loss of weight, maintaining leg strength and leg volume, and maintaining the integrity of muscle systems in general. . . . However,

in-flight exercise by no means offers complete protection. Cosmonauts Berezovoï and Lebedev returned from their recent 211-day flight aboard Salyut 7 in obviously debilitated condition. . . . Although they had exercised daily, their muscles were so flabby that they were barely able to walk for a week, and for several weeks afterwards required intensive rehabilitation.

Although in-flight exercise has been shown to benefit external muscles, the benefits to other physiological systems must be questioned. The Skylab exercise program did not deter decalcification or related problems of the skeletal system.

—Mary M. Connors, Albert A. Harrison, and Faren R. Akens, Living Aloft, *NASA Special Publication 483*

Astronauts Story Musgrave (left) and Donald Peterson evaluating handrail systems on the *Challenger* above Mexico.

The first *Skylab* Saturn rocket blasted off, carrying just the Space Station; the astronauts were to rendezvous with it on the next flight. Radar showed that the spacecraft had entered a proper orbit around the Earth. But some kind of serious damage had occurred. The two "wings" of solar-power cells hadn't spread, so there was a severe power shortage. Would it be safe to launch the astronauts?

After much debate, NASA launched them, figuring they'd have to return to Earth if the damage couldn't be repaired, killing the entire *Skylab* mission.

The astronauts' rocket matched orbits with *Skylab* and slowly approached it. Gradually, they could see that one whole wing of the Space Station had been torn off when an aerodynamic shroud failed in flight. And the other wing, which was supposed to automatically unfold straight out from the side of the lab, was still stuck closed, unable to generate power.

The spacecraft was not rotating properly, and as a result it was seriously overheating in the glaring sunlight. The main danger was that heat might cause toxic gases to be released from materials on board, poisoning the astronauts when they boarded it.

So when the astronauts rendezvoused with *Skylab,* they put a protective cover over it, to reflect the hot sunlight, allowing the craft to cool down. They released the stuck wing so that it unfolded and began to generate electricity. They cautiously entered the hot spacecraft, purified the air, and adapted the equipment to operate with the lower amount of power available.

Skylab had been designed to test the value of humans in space, but no one had expected the test to be so dramatic. That the astronauts were able to salvage *Skylab* after it had seemed hopelessly lost proved what the pro-astronaut people claimed: humans can adapt and fix things when they go wrong.

R2D2 can't completely replace man or woman in space. Yet.

HOUSTON, WE HAVE A PROBLEM

The crew's first EVA [extra-vehicular activity—space walk] was delayed again on 2 August [1973] by a faulty steering rocket that, for a while, threatened the entire mission. Apollo's reaction control system consisted of four independent sets of rockets spaced 90 degrees apart around the service module. Each set had four thrusters, hence the common designation, quad. Astronauts fired the rockets singly or in pairs to stabilize the spacecraft's position in orbit or to change velocity; the thrusters could also return the spacecraft to earth if the main service engine failed. It came as a surprise when quad B developed a leak on launch day—the reaction control rockets had been among Apollo's most reliable systems. Skylab procedures, however, provided for spacecraft operations with one quad shut down.

Surprise turned to alarm six days later when temperatures in quad D fell below normal limits. The drop triggered a master alarm, alerting Mission Control and waking the crew. At first the malfunction seemed minor, and the problem was not immediately connected with the first day's leak. Crewmen activated heaters in the reaction control system and turned to other duties. During the next hour, Mission Control received positive indications of a second leak: temperature and pressure in quad D dropped sharply and the astronauts reported a stream of sparklers outside their window, similar to the crystals they had seen the first day.

JSC [Johnson Space Center, Houston] engineers assumed the worst—that the two leaks represented a generic problem in the oxidizer portion of the reaction control system. . . . If this was true, the other rockets could soon fail. An oxidizer leak could also damage electrical circuits within the service module. Although quad D had lost less than 10% of its oxidizer, there was no telling how fast the leak might expand. The astronauts could maneuver the spacecraft with two quads, or perhaps even one, but it was a situation to avoid if possible. . . . Skylab's rescue capability, added three years earlier, suddenly looked like a good investment. According to Glynn Lunney, Houston's spacecraft manager, "if we did not have a rescue capability we would be . . . getting the spacecraft down as rapidly as we could."

At Kennedy, the news had an electrifying effect. Within three hours preparations for a rescue were under way. By eliminating subsystem tests at the Operations and Checkout Building, the spacecraft could be mated with its Saturn launch vehicle the following week. At the pad, storage lockers could be removed from the command module to make room from additional couches. Foregoing the traditional countdown demonstration test, the Launch Opera-

tions Office expected to have a vehicle ready in early September.

Tensions eased considerably when JSC engineers concluded that the two thrusters did not share a common problem. . . . JSC officials believed the two quads were still serviceable; if not, simulator operations indicated that the spacecraft could return safely without them. Kraft notified the crew that EVA would be delayed again, this time so that Mission Control could prepare procedures for reentry with two operational quads. He noted that rescue operations were under way as a matter of prudence, but that "we're proceeding as if we're going to have a nominal mission."

The leaking thrusters pointed up strengths and weaknesses in the Skylab operation. A subsequent investigation attributed the failure in quad D to loose fittings in the oxidizer lines which had gone undetected during two years of tests. When the crisis struck, NASA officials were not certain that the crew could deorbit with only one or two operating quads. Fortunately, Skylab's rescue capability meant that no decision had to be made immediately, and within a few hours the spacecraft's condition had been correctly assessed. The mission continued.

—*W. David Compton and Charles D. Benson,* Living and Working in Space, *NASA Special Publication 4208*

SIX

A Room with a View

SPACE STATIONS

There are three astronaut-oriented civilian space projects often hotly debated in the space community today. These are the Space Station, a lunar base, and a human expedition to Mars. In this chapter, we'll focus on the first.

Soon people will live permanently in space, able to watch with their own eyes as the Earth rolls by beneath them every hour and a half. This will happen whether or not the U.S. builds a space station, because the Soviets already have.

Few people think about it—to them, it's just another item in the news, forgotten with the day's scandals—but there are two Space Stations orbiting the Earth right now. Both of them are Russian: the *Soyuz* ("Union") *TM-2* and the *Mir* ("Peace"). The *Soyuz TM-2* is the outgrowth of the lengthy *Soyuz* series of spacecraft. Cosmonauts have spent up to eight months in the station, quite a bit more than the U.S. record of almost three months in *Skylab*. The *Mir* is a new, more advanced space station, and they plan to add at least four large modules to it. Currently it is 92 ft. long, and weighs 36 tons. It's clearly a step in the direction of a permanently occupied station, and perhaps the beginning of a space city.

The U.S. has plans to launch a space station, too. President Reagan gave the go-ahead, and NASA has been studying various proposals from aerospace companies, with the intention of launching sometime in the 1990's. Budget problems have plagued the Space Station incessantly, and recently the press publicized an internal memo in which

(a)

Proposed Space Station *Freedom* designs.
(a) McDonnell-Douglas (NASA's choice)
(b) Rockwell
(c) Boeing

several astronauts severely criticized the safety and comfort of the present designs. The delay in the space program caused by the *Challenger*
disaster will allow some of these problems to be resolved, but there
is a real possibility that they will delay it so much that a future administration may decide to kill the station altogether.

(b)

(c)

Tonight, I am directing NASA to develop a permanently manned space station, and to do it within a decade.

A space station will permit quantum leaps in our research in science, communications, and in metals and lifesaving medicines which can be manufactured only in space. We want our friends to help us meet these challenges and share in the benefits. NASA will invite other countries to participate so we can strengthen peace, build prosperity, and expand freedom for all who share our goals.

Just as the oceans opened up a new world for Clipper ships and Yankee traders, space holds enormous potential for commerce today. The market for space transportation could surpass our capacity to develop it. Companies interested in putting payloads into space must have ready access to private sector launch services. The Department of Transportation will help an expendable launch services industry to get off the ground. We will soon implement a number of executive initiatives, develop proposals to ease regulatory constraints, and, with NASA's help, promote private sector investment in space.

—President Ronald Reagan,
State of the Union Address, January 25, 1984

It may be surprising to the public, but space scientists have been among the most vocal of those opposed to building the Space Station. Many of these individuals feel that the money that goes into the human space program kills robotic projects. Certainly, in the past, the unmanned space program has often been shortchanged because of budget problems caused by the much more expensive piloted program. Scientist James Van Allen, who discovered Earth's radiation belts (named for him) with the first American spacecraft, has been a leader in challenging the use of piloted spacecraft. Such scientists hope to do away with the Space Station so that more funding will be available for Earth-based astronomy projects, robotic space probes, and other space programs that don't rely on astronauts.

On the other hand, scientists such as former astronaut Brian O'Leary are strong supporters. Ironically, O'Leary once attacked the human space program, even after being a part of it personally (his views of that time can be read in his autobiography, *The Making of an Ex-Astronaut*). But O'Leary eventually came to recognize the profound, long-range importance of humans living and working in space.

Similarly, the former leader of the American planetary space program, Bruce Murray, currently vice-president of The Planetary Society, says that without the piloted space program, there would be no unmanned space program (though he doesn't support the Station).

From "Space Science, Space Technology, and the Space Station," by James A. Van Allen. Copyright © 1986 by Scientific American, Inc. All rights reserved.

The major emphasis in recent years in space science has been on billion-dollar missions, such as the Space Telescope, the Galileo mission to Jupiter, the Viking landers on Mars and the Voyager probes. This trend also accounts in part for the reduction in scientific payloads; indeed, the Space Telescope and Galileo are the only major U.S. scientific spacecraft that have been or will be scheduled for launching in the years from 1983 through 1988. Such missions represent a tendency within space science toward ever greater complexity and sophistication, and they do have high merit. Unfortunately, however, like the large, manned space projects, they tend to squeeze out more flexible and much less expensive undertakings that historically have been highly productive. Smaller projects nurture space science on a broad, national basis and continue to have a potentially important role in our national program, but they are now nearly extinct.

In the meantime the European Space Agency, Japan and the U.S.S.R. are forging ahead with important scientific missions. The progressive loss of U.S. leadership in space science can be attributed, I believe, largely to our excessive emphasis on manned space flight and on vaguely perceived, poorly founded goals of a highly speculative nature. Given the current budgetary climate and a roughly constant level

> *of public support for civil space ventures, the development of a space station, if pursued as now projected, will seriously reduce the opportunities for advances in space science and in important applications of space technology in the coming decade.*
>
> —*James A. Van Allen, University of Iowa, 1986*

Still, scientists have legitimate complaints about the Space Station. For instance, astronomical telescopes are so fanatically precise that the slightest disturbance on board the craft can blur the image hopelessly. Rockets arriving or departing from a space station, or even astronauts moving around inside, can ruin photography. Rocket fumes can contaminate experiments. The solution is to keep such instruments detached from the station, and even to put them in separate orbits on platforms that could be visited and serviced by astronauts. Then the question becomes, do you need a permanent station just for occasional visits? But there's much more to the Space Station than astronomy.

What Do You Do with a Space Station?

Instead of focusing on the problems, let's take a look at what NASA plans to do with the Space Station if it is built. The station would orbit a couple of hundred miles above Earth, easily accessible to the Space Shuttle. It would serve as a place to check satellites before launching them into their final orbits, whether Earth orbiters or interplanetary spacecraft; and it could be used as a fuel base to refuel spacecraft, both piloted and robotic.

Certain materials and medicines that are difficult or impossible to manufacture on Earth could be made in the station's zero-gravity (zero-gee) environment. Its laboratory would also make it possible to perform long-lasting experiments for commercial use of space. And it's a great place to keep equipment with which to repair other satellites when they break down.

The Earth can be monitored, its weather and agriculture followed, and a human observer on board may spot changes that would be overlooked by preprogrammed satellites.

Some of this work can, it is true, be done by automatic machines. Small factories, for example, have been designed to produce medicines in space, to be checked by astronauts at intervals of weeks or months. But the machines must be frequently supplied with raw materials, and they need routine maintenance, as well as occasional repair. All these functions require astronauts. Even robots in Earth's factories require daily human supervision or they quickly break down.

Before we get carried away by the thought of robots replacing astronauts in space, we should look at our earthly experiences with these machines. A good example is the U.S. Army's Autonomous Land Vehicle (ALV). ALV, designed by the Martin Marietta Corp., is supposed to be a robotic vehicle that can travel into enemy lines and return. Currently, even with its TV camera and laser scanner, it can't go faster than 6 mph on a rough, winding road, nor can it handle the change from road to dirt, or even a change in the weather.

The problem is that the human mind is a fantastically better pattern-recognizer than any computer yet built. Astronauts need not line up at the unemployment office yet.

What else can you do up there? A space station is the ideal place to test possible new spacecraft. New types of propulsion have been developed in the laboratory that need to be tested in space. Two of them are especially promising as inexpensive, efficient ways to send space probes around the solar system: ion drives and solar sails.

Ion drives are a special type of rocket that accelerates ions—charged

---

Man must have bread and butter, but he must also have something to lift his heart. This program [Skylab] is clean. We are not spending the money to kill people. We are not harming the environment. We are helping the spirit of man. We are unlocking secrets billions of years old.

—Farouk El Baz, NASA geologist, 1973

atoms—by electricity, which can be generated by solar cells. Solar sails would use great umbrellas of thin plastic to catch the gentle pressure of sunlight as a free source of power. With it, a spacecraft may sail on sunbeams between the planets like a clipper ship on the Earth's high seas. A space station would be a good place to perfect such designs.

A space station is the only place where we can study the effects of long-term weightlessness on human beings, information that is crucial to planning a piloted Mars mission. Serious discussions are underway among the U.S., the U.S.S.R., and the European Space Agency about such a joint expedition to Mars. We need to find out how to combat the weakening effects of a year's travel in zero-gravity, so the astronauts could survive the rigors of a landing on Mars. If they can't handle it, then we'll have to design more complex spinning spacecraft to simulate gravity.

And a space station would be a place where ordinary citizens could come and visit, to experience life in space. Writers, TV reporters, poets, politicians, teachers, and other citizens who have experienced only the familiar conditions of Earth would get a new perspective on our civilization by going into orbit. Virtually every astronaut has come back to Earth a changed person, and the orbital experience could not fail to enhance the perspective of the non-astronaut visitor.

Orbital Alchemy

The alchemists of old wanted to convert common lead into precious gold. Zero-gravity enables us to perform even greater magic: the conversion of ordinary metals into something rarer than gold—alloys that could never be made on Earth, new metals with properties we're just beginning to discover.

On Earth, many liquids cannot mix—oil and water, for instance. These are known as immiscible liquids; the lighter one floats on top of the denser. In space there is no "on top." As a result, the light and heavy can be mixed. Under such conditions, we can blend metals, such as aluminum and iron, that do not mix even when molten on the Earth.

We don't yet know what value these alloys may have to the future of technology. But the history of technology is, to a great extent, based on the development of new materials. Much of the prehistory of the human race was called the Stone Age with good reason; stone was the only major material out of which to make tools, other than the soft wood and bone used before that.

Then, for many centuries, our ancestors labored in the Bronze Age using that mixture of copper and tin. Next, they discovered how to purify iron and make crude steel, and the Iron Age was born, offering new tools, new plows, new abilities to manipulate the environment.

Zero-gravity alloys may give us whole new substances with which to revolutionize our civilization. There may be alloys far stronger but lighter than anything we now use. Perhaps the very buildings we make will one day be made from zero-gee substances.

Then there is the strange phenomenon of superconductivity. To a great extent, our civilization today runs on electricity, and that electricity has to be transmitted from point to point, usually by wires. At ordinary temperatures, any conductor has some resistance to electricity. The resistance acts like friction and causes some of the electricity to be wasted as heat in the wires.

Early in this century, a Dutch scientist, Kammerlingh Onnes, discovered that if you cool certain metals to very, very low temperatures, their resistance disappears.

It is rare in science to find a phenomenon of absolute perfection. The materials of our universe are plagued by imperfections that create friction and heat and hence waste energy. But the resistance of a superconductor is not just very small; it's not even very, very, very small. It's *exactly zero!* If you start a current circulating in a closed loop of superconducting wire, that current will travel endlessly for years without undergoing the slightest decay.

It took decades to understand this phenomenon. It wasn't until one of the inventors of the transistor, John Bardeen, and his colleagues applied the quantum theory of matter to the problem that they were able to understand the strange interactions between the superconducting electrons that allow currents to flow eternally. For this, they won the Nobel Prize.

The reason superconductivity hasn't had much practical use, despite our knowledge of it for so many years, is that materials have to be cooled down to nearly the coldest temperature in the universe, absolute zero. Temperature is the movement of molecules, and at absolute zero—456 degrees Fahrenheit below zero—all motion stops.

Scientists have been trying for years to find a substance that would be a superconductor at room temperature, and recent breakthroughs suggest this is possible. If they did, it could revolutionize electrical technology. Our constant waste of energy could be reduced. Huge power transmission lines could be replaced with small ones. Large motors and generators could be made small. Electric cars might become practical, reducing our dependence on limited, polluting fossil fuels.

Some of the zero-gravity alloys produced in space have turned out to be superconductors. It may be that, with further research, we will find a way to produce in zero-gravity a true room-temperature superconductor.

And that is just one possibility. We will probably find other substances that are as revolutionary as rubber, plastic, aluminum, and silicon have been in the last century. Only the modern orbital alchemist, floating in zero-gee, living there day after day, is likely to see the kind of *unexpected, unpredictable* quirks that have often sparked breakthroughs in the history of science.

Already, experiments on board the Space Shuttle have shown the complex behavior of liquids in zero-gravity. There, a drop of water tends to become a sphere. An experiment that manipulated liquid drops by sound waves showed how liquids behave under unearthly conditions. Studies of the results are being used to test theories of liquid behavior that could never be conclusively tested on Earth before; they could ultimately lead both to fundamental advances in our understanding of familiar fluids and to practical techniques for handling molten metals in space.

The most practical value of astronaut activities in space operations for the near future may be medical. Johnson & Johnson and McDonnell-Douglas corporations built a miniature space-manufacturing laboratory that has been carried on the Space Shuttle. What they have found so far is that some chemicals that are extremely difficult to separate on Earth, due to gravity, can be made much more efficiently in space. Medicines that are currently too expensive to manufacture on Earth may be practical to manufacture in space. Very soon, we expect that lives will be saved here on Earth by the medicines made in space.

However, Johnson & Johnson has dropped out of the project for a reason that shows an unexpected form of competition to space manufacturing: Biotechnology came up with a completely different technique for producing the chemical they were primarily interested in.

The first practical manufactured product from space is a modest one: microscopic plastic spheres used for calibration in industry and medicine. It is difficult to make them uniform on Earth, because gravity distorts the droplets while they are liquid. But in zero-gravity, a droplet forms a near-perfect sphere. You can now buy a box of these little made-in-space balls off the shelf.

The U.S. Space Station

The story of the American Space Station really began with President Reagan's election. Like his Democratic predecessor, Jimmy Carter, President Reagan was not a space buff when he came to office. Initially he kept cutting back on various space programs, as Carter had. But at some point, Reagan became a space enthusiast, eventually coming to the decision that the U.S. should have a permanent space station. He gave NASA a fixed budget of $8 billion with which to build it.

This figure was destined to cause trouble, because the Space Station that NASA wanted to build would have cost substantially more. Trying to make a cut-rate space station do everything that scientists and all the potential users wanted was very difficult.

The more NASA studied the Space Station, the more they had to cut back on its size and scope. Initially, for example, they hoped to have seven modules in which the astronauts would live and perform experiments, but now they have had to cut this down to four. Even with these cutbacks, the cost estimates have been soaring to almost double.

Then, following the *Challenger* disaster, astronauts became much more critical than they'd been for a long time. A memo circulated among them saying that the station was poorly designed and would be difficult for astronauts to live in for long periods. This is the memo that leaked to the press, forcing NASA seriously to reexamine the original design.

It brought to mind the early days of the astronaut program, when astronauts objected to the initial designs of the *Mercury* and *Gemini* space capsules, on the grounds that they were controlled by Earth, so the astronaut was really just a passenger, not a pilot. "Spam in a can" was the common expression among astronauts in those days.

They pressured the design engineers to give pilots more control over their spacecraft. NASA eventually gave in, to the benefit of the spacecraft. Later, when problems arose and a spacecraft was on the verge of disaster, the astronauts had enough control so they could fix the problems well enough to survive and return home safely.

Life in Space

One of the biggest problems of living in space is that life is too easy in zero-gravity. We are born on a planet where we constantly have to fight gravity, and our muscles and bones are used to it. In space, without that everyday stress, the body gets weak, and calcium and minerals begin to leave the bones, making them soft and fragile. Astronauts who have spent months in space sometimes resemble hospital patients who have spent a similar amount of time in bed.

Another problem, probably a combination of our psychology and

New knowledge was obtained concerning astronauts' responses in the zero-g environment. There was reasonable stabilization of their physical reactions after about the 40th day in space. Upon return to Earth, the crew of the second manned Skylab mission (59 days) readapted more quickly than did those on the first mission even though they had been in space over twice as long. The results were encouraging. On the basis of this experience, tentative approval was granted to extend the third manned mission to 84 days. It is significant that the members of the 84-day mission returned in even better physical condition than did the crews of the previous missions. Obviously, travel in space for prolonged periods is feasible.

—Space Flight Research *(NASA pamphlet)*

physiology, is that astronauts often initially find it disturbing to live in zero-gravity, with no up and no down. In fact, many astronauts get spacesick—they feel dizzy, sometimes vomit, and exhibit other symptoms similar to airsickness. We don't know what causes this, nor are there any tests that can predict who will suffer from spacesickness. Most astronauts find it helps to arbitrarily call one side of the spaceship "up" and the other "down." Others find it exhilarating to ignore the concept of up and down.

Spacesickness usually goes away before long, and many learn to enjoy zero-gravity, which resembles, after all, the natural state of our ancestors, who came from the seas.

Internationalization

The Space Station has stimulated considerable interest among our allies. The Western Europeans are afraid of falling behind the U.S., in much the same way the U.S. fears competition with Japan. The Euro-

Your nation's historic decision to build a permanent Space Station has created great interest on all sides. It is another manifestation of your nation's capacity to create, to adapt and renew, which has shown itself in many ways throughout your history. It will favour the development of new space techniques.

Following President Reagan's invitation to the friends and allies of the United States to participate in the Space Station program, and based on our own scientific and technological capabilities, we are considering ways to join forces with you for the advancement of this project and at the same time challenge our own imaginations and skills. We hope that the Space Station will enhance the extensive transatlantic cooperation which has served our peoples on both continents so well in the past.

—E. Quistgaard, Director General, European Space Agency

peans and Japanese know that space technology has been one of the greatest spurs to American technology as a whole. So it is not surprising that both the Japanese and the Western Europeans have offered to participate in the Space Station, to the tune of three billion dollars, to make sure they have access to the station itself and to any associated technology. Their contribution, of course, would reduce the cost to the American taxpayer.

The Future of the Space Station

Scientists know we have to explore new frontiers in order to advance the sciences. Typically, their greatest difficulty is convincing the research funders to support basic research, which often seems to have no practical application, even though such research in the past has usually led to breakthroughs that benefit us all on an everyday basis. It's ironic that many scientists—those who know the value of raw research—are critical of the Space Station, where so much fundamental experimentation could be conducted.

Simulating the repair of a Space Station module in the NASA Marshall Space Flight Center's Neutral Buoyancy Facility.

In time, the Space Station may cause us to adjust our point of view so that we will look on these Earth-orbiting regions as integral parts of the Earth, like newly-found continents or oceans.

By then—perhaps in the late 1990s—commercial manufacturing activities in orbit may have outgrown the original Space Station. Modules especially designed to meet the precise needs of these commercial enterprises will have been added. Similarly, scientific research and applications needs will have expanded so that dedicated modules will have been added to satisfy these requirements.

By that time the Space Station could be in the process of becoming a staging and launch base for manned voyages to the Moon, Mars and asteroids. Expeditions to these remote destinations can best be started from the Space Station rather than from Earth. By assembling the spacecraft in orbit and launching it there, it will not need to be equipped for a strenuous passage through the Earth's atmosphere. Instead, it can be constructed entirely for use in the vacuum of space with resulting economies and design advantages.

The Space Station with its added specialized modules for new functions and activities can become an important legacy of this generation to the 21st century.

—Walter Froehlich, Space Station,
NASA Educational Publication 213

The history of science and technology suggests that as soon as a new niche in the technological environment is found, practical engineers and inventors find totally unexpected ways to fill that niche. It was the study of the use of humble hot water that gave rise to the steam engine, which powered the Industrial Revolution. The Jacquard loom of the 19th-century weaving industry gave birth to the computer punch-card. The discovery of the curious electrical properties of almost-but-not-quite-pure exotic materials gave us the silicon chip. Study of the laws of electricity and magnetism gave us the theory of relativity.

I'm sure similar results will follow from space research. Medical care has already been revolutionized by sensors and monitoring systems developed for astronauts. (The great science-fiction writer Robert Heinlein said he owed his life—appropriately—to this space technology, which saved him when he was gravely ill.)

If we have people living up there day after day in zero-gravity, able to modify their research as they get new results, devising new experiments on the spot that don't have to be planned and approved years in advance, they will undoubtedly make discoveries we cannot guess at now. In the long run, a space station could pay for itself in both the practical and scientific fruits of its research.

The Space Station would also be the place to start testing what may be the main power source of the 21st century: solar-powered satellites. As Peter Glaser has shown, solar energy can be efficiently converted to electricity in space and transmitted to Earth as a microwave beam.

This, physicist Gerard O'Neill believes, may be the biggest economic benefit of space for the foreseeable future, and could replace most of civilization's dependence on fossil fuels and nuclear power. Many of the techniques he and his colleagues propose need to be tested in space, and the Space Station is the ideal place.

It's very likely that, within 25 years, we will see practical solar-power satellites in orbit, exporting energy to Earth.

There's a vastly different universe above our heads, just a hundred miles up. Until now, humans in space have been explorers, like the first men who ventured across the oceans to the New World. But now we are entering an era in which the astronaut is the equivalent of a truckdriver.

Without a permanent presence in space, we would be like a man living on a seashore who occasionally puts his toes in the water and thinks he knows all about the ocean. He never learns about the fascinating creatures that live in the sea and the joys of a boat ride across the waves or the uses of the oceans for food and natural resources. And he never knows how much he has missed.

SEVEN

The Man and Woman in the Moon

A LUNAR BASE

Many a night we can see overhead a beautiful world filled with riches just waiting for us. It's been howled at by wolves and admired by lovers, and it may be the most important piece of real estate in the future of civilization: the Moon.

The U.S., Europe, and U.S.S.R. have all been designing different spacecraft to investigate an exciting possibility: There may be ice in the poles of the Moon. This would be extremely important because so far no water has been found on the Moon in any form, and water would make colonization far easier.

The Moon is one-quarter the diameter of Earth, larger than the planet Pluto (which may actually be a former moon of Neptune). Since ordinary weathering does not disturb the surface of the Moon, it's an ideal laboratory in which we can learn more about the formation of the Earth. The rocks that the *Apollo* astronauts brought back have given us great insights into the history of the Earth and Moon, but they still have not answered the question of how the Moon formed. Is the Moon a piece of the Earth? Or did it form separately, like one of the other planets? Was it then captured by Earth's gravity?

The low gravity of the Moon—one-sixth of Earth's—makes it a good place to find out if such a small attraction can prevent the weakening effects of free-fall on astronauts who spend months in space—such as those who will one day live in permanent space stations or travel to Mars.

The back of the Moon is probably the best place to set up observatories to see the rest of the universe. To astronomers, the Earth itself is growing increasingly polluted with interference. Optical astronomers have to combat smog, haze, clouds and the atmospheric reflection of city lights, all of which degrade the capability of even such great telescopes as Palomar. To radio astronomers, the skies are filled with endless signals produced by satellites, microwave dishes and other transmitters and noisemakers that mask the emissions of natural objects in the universe, such as galaxies, quasars, and pulsars. It is also becoming harder to detect any possible signals of other civilizations in the galaxy, if they exist.

These problems can be reduced by putting telescopes into orbit around Earth, but the best place of all for an observatory would be the far side of the Moon, shielded from such interference, since that side faces permanently away from Earth.

But the ultimate value of the Moon to humanity is probably its enormous stockpile of natural resources. The *Apollo* astronauts found that the Moon contains enormous quantities of such strategically valuable materials as titanium, a space-age metal. Lunar rocks contain oxygen in the form of oxides that could be used to supply a settlement on the Moon.

The very dust on the Moon, I suspect, may turn out to be one of its most valuable resources. There are numerous oceans of moondust produced over billions of years by meteorites hitting the surface. This dust is a sample of all the minerals on the lunar surface, and it could be mined easily with a scoop. In the vacuum of the lunar surface, it could be separated and chemically refined to extract its minerals.

The airless, low-gravity conditons on the Moon make it prime real estate for long-term industrial development. It takes only one-twentieth the amount of energy to move an object from the surface of the Moon that it takes to move the same object from the surface of the Earth. Because of that difference, and because of the lack of an atmosphere, it's much cheaper and easier to get resources off the Moon and into space than it is to launch them from Earth.

THE INTERLUNE CONCEPT

. . . international experience has provided a successful model of a high-technology management system that meets the legal, operational, and self-interest constraints attendant to international operations in space. That model is INTEL-SAT, a user-based management system that works to coordinate operation of international communications satellites. . . .

In this paper, a version of the INTELSAT model is suggested as being especially appropriate for the international management of a lunar base. It is submitted here that "INTERLUNE," as the organization is termed, will aptly satisfy the aforementioned legal constraints, as well as hold consistent with the principles of free enterprise that are shared by the world's democracies. More importantly, INTERLUNE would bring into the management of a lunar base those states and other interests that evince the greatest motivations for ensuring successful implemenation of that managerial system. INTERLUNE does not require that sovereignty be given up in space; it does not require that free-enterprise opportunities be abandoned in space; it merely requires that sovereignty and opportunity be shared.

Technological advancements have produced a trend toward recognizing a "common heritage of mankind" in certain international resources. This trend is most apparent in negotiations regarding the resources of the sea and outer space. It indicates a general realization that states possess common interests in sharing benefits from the exploitation and environmentally sound use of those resources.

The Moon can become a common heritage resource for mankind. However, without a feasible administrative system and a peaceful management environment, lunar opportunities will remain unavailable and moribund. An institutional arrangement should be possible that would vest operation and control of lunar bases in an organization comprised of states that actively participate in creating such bases, in association with users of the bases or investors in their operations. Such states and related entities would be united by a common bond of policy and purposes focused on the technical and financial success of the enterprise. . . .

> *The conceptual advantages of a regional organization such as INTERLUNE could be realized only if the actual institutional structure were designed to provide an equitable system for various interests to exert influence and control, as well as to furnish efficient and proper management of the base.*
>
> *The main functioning bodies INTERLUNE is comprised of are the Assembly of Parties, The Board of Governors, the Board of Users and Investors and the Director's Office.*
>
> —Christopher C. Joyner, George Washington University, and Harrison Schmitt, Apollo 17 *astronaut and former U.S. Senator*

For any extensive development of the solar system, such as building large space stations and major space colonies, the most economical way is probably to mine the Moon, although we might find that asteroids, mountains of rock that lie largely between the orbits of Mars and Jupiter, would be an economically competitive resource for such construction.

We must explore the Moon further to answer questions raised by the *Apollo* samples, and to find out what other resources exist. Comets and meteorites that have crashed onto the Moon in the past have doubtless deposited there large quantities of iron, nickel, and carbon compounds, all of which would be useful materials for colonists.

The apparent absence of water is probably the greatest single obstacle to inhabiting the Moon. Water is not only essential to life, but it contains oxygen that can be easily separated from the hydrogen by electricity to produce air for breathing, as well as oxidizer for rocket fuel. Oxygen is heavy, so to bring it from the Earth would cost a lot of fuel. If we can get our oxygen from the Moon, the cost of living in space would be greatly reduced. It would be easier to extract oxygen from water than from minerals.

Hydrogen is rare on the Moon, but if there is water, hydrogen can be separated from it, providing one of the best rocket fuels.

Most ice, if it once existed on the surface of the Moon, has long since been evaporated into space by the heat of the Sun. However, at the poles, it could be cold and shaded enough to have lasted to the present time. The U.S., Europe, and the U.S.S.R. are all working on designs for robotic lunar orbiters that could detect water if it exists there. So important is this mission that at least one of those spacecraft will probably be launched in the next decade.

Moonbase

There is a good chance that a permanent scientific base, much like those in Antarctica, will be established on the Moon within the next 25 years. Geologists then could further explore, at their leisure, the puzzling issues raised by *Apollo*. Engineers could test the feasibility of extracting lunar resources. We could get started on such a base in the next ten years, if there is enough public support.

The Moon has a surface area roughly equal to one of Earth's continents. Probably every metal needed by civilization is there in enough abundance to supply the Earth for the foreseeable future. The question is whether this world of resources can be made available to us at a reasonable price.

When we think of going into space, we think of the huge rockets required to get us off the Earth. Many people assume that it would require a similarly huge project to get any resources off the Moon and into space. But, as pointed out earlier, that effort would be much less on the Moon than on Earth. If we were to mine the Moon and process the materials there, we could shoot them off the surface of the Moon with relative ease, perhaps using electromagnetic catapults, something like those that launch airplanes off airplane carriers.

The materials could be caught by factories orbiting the Earth and used to expand them, as well as to build larger space colonies, eventually perhaps exporting further-refined lunar materials to Earth.

If we are to build factories orbiting the Earth or the Moon, resources can probably be more inexpensively shipped from the Moon out into orbit than from the Earth. At first, of course, the space construction would have to be supplied from the Earth, but later on, we'd be able to supply most of the needs of the space dwellers from lunar resources.

The catapults themselves would mean great savings, since they would eliminate the need for chemical rocket fuel. Catapults can be driven by electromagnets, powered by solar energy or other sources. In the laboratory, scale-model catapults of the type needed to get materials off the surface of the Moon have already been built.

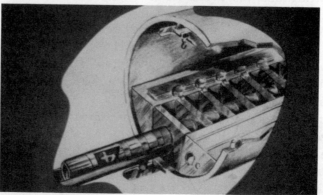

Designs developed by Electromagnetic Launch Research, Inc., for launch of space materials. This may be the most economical way to get matter from the Moon into space.

AUSTRALIAN PARALLEL

Australia was very far from England, and, as the first parties discovered, a continent less blessed with Nature's bounty than supporters of the settlement had led the government to believe. It was a dry land, subject to frequent drought, and British crops did poorly. . . . [Free settlers] and the convicts who had served out their terms found ways to make livings and gradually to build a self-supporting community. All that happened in less than a lifetime, and by 1840 Australia was, for practical purposes, standing on its own feet.

What does the Australian experience have to teach us, as we contemplate a return to the Moon? Australia is not the Moon; it it a terrestrial environment far more benign than our airless satellite, but there are similarities. In broad brush a few pertinent features stand out. Living on the Moon is going to be a new and initially difficult experience. The lunar base will start small and be very dependent on Earth. The first "settlers" will have to be technically trained: astronauts, engineers, and scientists. . . .

If the lunar settlement is to grow and eventually become self-supporting, some helpful features can be built into this social experiment from the beginning.

First, the stated purpose of the lunar base must be the eventual establishment of a self-supporting lunar settlement. If that purpose is clear from the beginning, then the inevitable transition from a tightly administered scientific base to a more open community may come more easily. If, on the other hand, we say that we are going for purely technical reasons, then the interests of entrepreneurs and private settlers (on whom ultimate success may well depend) might not receive proper attention.

Second, reducing the cost of transportation must be a main concern, although we anticipate that these costs will remain high for a period of time. During that initial stage subsidies of public and private ventures will be required.

Third, there should be emphasis on the development and encouragement of profitable enterprises through technology transfer, tax incentives, and the development and maintenance of basic services.

Fourth, there should be mechanisms established for the

orderly transfer of control of lunar operations to the settlers as the population grows—mechanisms analogous to the Ordinance of 1787 in the United States and those used by Britain to create independent nations such as Australia, New Zealand, and Canada.

There will be conflicting interests, shifting purposes, short-sightedness, greed, and mistakes. The lunar base is going to be expensive and will remain so for longer than some of us would like. If we give careful thought to the design of the experiment now, however, considering the human as well as the technical aspects, and keep our eyes open both for the pitfalls and the opportunities that will come along later, perhaps self-sufficiency of the lunar settlement will come quickly.

Finally, we should remember that Britain supported the Australian settlement long enough for it to succeed. Let us hope that we can do the same for the lunar enterprise.

—Eric M. Jones, Los Alamos National Laboratory,
and Ben R. Finney, University of Hawaii

Everyday Lunar Life

Life on the Moon would obviously be a dramatic change from life on Earth. Beyond psychological effects, lunar colonists would face certain advantages and problems. A big plus (over a colony elsewhere in space) would be the Moon's natural protection from solar flares. Occasionally, the sun flares up, releasing hazardous radiation that could endanger the lives of anyone outside the protection of Earth's atmosphere. Dwellers of space colonies in orbit would have to have the equivalent of a midwesterner's storm cellar—a place to go and sit out the radiation storm in comfort behind Moon-rock walls. On the Moon, it would simply be a matter of going into an underground dwelling.

One of the problems of living on the Moon is that a single day-night cycle is about a month long. Since the Moon keeps the same face toward the Earth constantly, and it takes about a month to go around the Earth completely, the Moon sees the Sun for two weeks and then sees the blackness of space for two weeks, except at the poles.

This makes energy supply a possible problem for lunar dwellers.

LUNAR INDUSTRIALIZATION

Selenopolis—the city-state of lunar civilization and the lunar biosphere—will be a network of enclosures gradually expanding to cover many square miles of the lunar surface, and some parts of the subsurface.

The enclosures comprise sections that are several miles long, with interior dimensions of 3200 feet across at floor level, and 1600 feet high to the center of a curved ceiling. The sections are joined at nodal points that serve as power, supply, and climatic control centers.

Selenopolis embodies urban, rural, agricultural, industrial, and resort areas. Each section is separated from the other by a solid but transparent "curtain," because each has a different Earth-like climate and season. . . .

Resort areas can include a winter section with snow, a subsurface lake for boating, a "sunbelt" with "lunar desert" views from a clubhouse that also overlooks the Alan B. Shepard low-gravity golf course, etc. Selenians may also enjoy the Krafft A. Ehricke rotating swimming pool. . . .

Selenians will not be bound to their biological environmental niches. In comfortable space vehicles, or in transport vehicles with interior shirtsleeve environments, they can tour the coasts of mare, the mountains, the cliffs of the southern highlands, the province of large craters stretching from the eastern coast of the Mare Numium to the South Pole, and more.

. . . Selenopolis is open-ended, growing with its population and advancing technologies. In principle, the overall complex could eventually house many hundred million people. Such a large complex is never completed, just as development of a continent is never completed. Like the giant cathedrals of the Middle Ages, Selenopolis will be the work of many generations.

—Krafft Ehricke, late NASA scientist

Solar energy would work well for the two weeks when the Sun is in the sky. But during the long night, energy would have to come from another source, such as fuel cells.

A fuel cell is a type of storage battery developed for the space program. It combines oxygen and hydrogen to form water, a reaction that releases energy. On Earth, when you combine hydrogen and oxygen, they usually burn or explode, and, in fact, this combination is the main source of rocket thrust in the Space Shuttle when it takes off. However, the gases can be made to combine inside a fuel cell so that instead of releasing heat, they produce electricity. One extra advantage here is that the process automatically produces water as well. Fuel cells are a clean, non-polluting way of getting two essential supplies— electricity and water.

One way lunar dwellers could use fuel cells would be to take advantage of the reversibility of the whole process. During the long lunar day, they could use solar electricity to break down water into hydrogen and oxygen, which would be stored. Then, during the night, the hydrogen and oxygen could be slowly recombined in the fuel cells to produce the energy needed to run everything when solar energy was not available. This process could be repeated endlessly: fuel from sunbeams. One day, we may use such fuel cells here on Earth, for transportation energy.

Twelve men have now walked on the Moon. Six widely different sites have been sampled. Orbiting satellites have mapped every place there except its poles. We know more about the Moon than any other heavenly body.

We haven't been back since 1972, when astronauts Cernan, Evans, and Schmitt visited the Moon for the last time, witnessed by millions of people, among them one Charlie Smith, a former slave, born eighteen years before the start of the Civil War; his life reflects the pace of change in our age.

A lot of people would like to go back to the Moon: scientists, astronauts, and ordinary folks who would like, at least once in their lives, to walk on another world. All the major problems were solved in the 1960's. With enough public support, we can return, to begin the first settlement of another world.

EIGHT

Who Stole the Canals?

MARS

Mars is in many ways the most fascinating place in the solar system, and also the pleasantest planet outside Earth for humans to live. It will be one of the highest-priority targets for the robotic space program in the foreseeable future, and will almost certainly be the first planet beyond Earth that humans set foot on.

The Soviets have sent a probe to Mars, and NASA has plans to do so in the near future. The Soviet probe was to study the tiny moon Phobos, but died soon after reaching Mars in 1989. Phobos is a mountain of rock that was probably captured from the asteroid belt. About

> The question, "why explore?" pertains less to the Viking 1 expedition in particular than to the nature of the human mind in general. We are here to consider not just the phenomenon of a journey to Mars but the phenomenon of intelligence. The fact that we can conceive of the inconceivable, and comprehend the incomprehensible, is perhaps the highest exercise of the human brain, symbolized so dramatically by the exploration of Mars.
>
> —Norman Cousins, author, July 2, 1976

JPL mathematician/artist Edward Belbruno's vision of a human expedition to Mars.

fifteen miles in diameter, it's an irregular, cratered moon with strange cracks in it whose cause is not completely understood. It may also contain resources that could be mined for supplies by future colonists or space travelers.

The Russian spacecraft had unique probes on it, beams of particles and laser light to vaporize tiny bits of the surface of Phobos and analyze their chemical composition. The American mission, called *Mars Observer*, has been approved by Congress, and is being designed to study the surface and weather of Mars. Its launch is tentatively scheduled for 1992.

There have been many discussions between Soviet and American scientists to coordinate their efforts, so that each mission will help the other and maximize the amount of information we get back. For the most part, there has been excellent cooperation in the sharing of information received from robotic exploration of space, which has not usually been the case with piloted spacecraft.

In addition to their separate missions, the two superpowers have been intensely discussing the possibility of a joint project to send a robotic spacecraft to Mars that would probe the surface and then shoot a sample of the Martian soil back to Earth for analysis. This *Mars Sample Return Mission* would help us answer many of the puzzles unearthed by the *Viking* landers, including a definite answer to the question of why two of the lander experiments seemed to detect life (probably false alarms). In the present political climate, this mission has a good chance to fly, perhaps in the 1990's.

And serious discussions are underway between the two nations to explore the possibility of sending humans to Mars, which could happen within the next 25 years.

Mechanical Martians

Mars, historically, is the planet of canals and Martians. By the end of the 19th century, the notion that there were intelligent beings on Mars had become so widespread that a Madame Guzmann in Paris announced a bequest of 100,000 francs as a prize to the first person successful in communicating with beings of another world *other than the planet Mars*. The French Academy was reluctant to accept this bequest, but they finally did so, possibly because her will included the condition that every five years, if the award had not been given, the interest would be used to support astronomy. The academy finally accepted it with a quotation from Montaigne: "It is a stupid presumption to condemn as false all that which may not appear likely to us. There is no greater madness in the world than to reduce everything to the measure of our capacity and competence."

When *Mariner 4* flew by Mars in 1965 and took the first close-up pictures of that world, it shattered our dreams and Madame Guzmann's presumptions. It revealed a world of large craters, moonlike in their appearance. According to these early pictures, there were no canals, no atmosphere, and no signs of life anywhere. *Mariner 4* appeared to have destroyed forever the hopes of many scientists that we would find life on Mars.

The next spacecraft in the *Mariner* series returned better and better pictures of Mars, and soon we discovered that Mars did have enormous riverbeds. Suddenly, the possibility of life became real again.

Everything we know about life on Earth suggests that water is essential to all living creatures. It appears that the evolution of life starts in the oceans, because water permits numerous chemicals to dissolve

Tonight, as we contemplate Mars, I feel as if I were standing on a threshold of immense dimension. All my life I have followed the explorations of Mars intellectually, philosophically, imaginatively. It is a planet which has special connotations. I cannot recall anyone ever having been as interested as we are in Jupiter or Saturn or Pluto. Mars has played a special role in our lives, because of the literary and philosophical speculations that have centered upon it. I have always known Mars.

But to be here tonight, to have seen that remarkable series of photographs which has come from that remote planet, and to realize what a weight of information they are bringing, what a freight of imagination and possible solution, is a moment of such excitement for me that I can hardly describe it. If the photographs I have seen do indeed show riverine action—I mean those marks which look like possible river terracing or the benchmarks customarily made by rivers—then I, for one, will have to admit that a major segment of my inherited knowledge has been shattered. Much of what I have believed about space will have to be revised, for we will now have in Mars a planet which once had a liquid component, which means that it had a substantial atmosphere, which means that it once had illimitable possibilities. Imagine living the days when a discovery of such fundamental significance was possible!

The Moon never caused me much trouble. I had to revise few of my concepts. After all, getting there was merely a technical problem. Scientists had already taught me as much about the Moon as I needed to know. It was a minor appendage attached to the Earth; it was egocentric. But when you move out to a planet which is a creation comparable to our own and which has similar propensities and possibilities, you are moving into a whole new orbit of speculation. The realization that in these very days, we are getting information from the threshold of our particular galaxy, . . . information which we can then apply to the billionth galaxy in farthest space, is to me an overwhelming experience. If subsequent photographs do produce evidences of riverine action, then we are faced with the question: Why did the water leave? What caused the great change? Is such change inevitable in all such successions?

> *What does such evidence mean concerning life on other comparable planets, the billions upon billions of other stars that are in this galaxy alone and the billions of galaxies beyond them?*
>
> —*James Michener, author, July 2, 1976, two weeks before the first* Viking *landing on Mars*

and react with one another. It is conceivable that there are other liquids in which life could arise, but water is the only one biologists agree has this ability.

Liquid water cannot exist on the surface of Mars as it is today. The pressure of the Martian atmosphere, about 1 percent of the Earth's pressure, is too low to permit liquid water to remain on Mars for very long. It would boil away quickly, despite the cold temperatures.

However, the presence of those huge riverbeds on Mars indicates that sometime in the Martian past there must have been liquid water. There must have been a thicker atmosphere, one that allowed pressure great enough for liquid water to exist. Where did that atmosphere go? Some of it certainly evaporated into space, since the Martian gravity is only about one-third that of the Earth's and hence cannot hold its atmosphere as well as Earth does.

Mars also has large, bright ice caps, which can even be seen from the Earth with a good telescope. Frozen water in the ice could once have supported life on Mars and may be used in the future by space colonists.

Recent re-analysis of *Viking* photographs of the surface of Mars has found features that some geologists believe are signs of more water underneath the surface than was previously thought to exist. They have also observed what seem to be shorelines of ancient seas.

By far the most ambitious space projects to date involving another planet were the *Viking* Mars landers. Two identical craft were built, each including an orbiter and a lander. The *Vikings* reached Mars in the American bicentennial year, 1976. NASA had, in fact, hoped to put the first one down on Mars on July 4th, but the evaluation of possible landing sites took longer than expected. They were not able to land until July 20th, which turned out to be the anniversary of the first *Apollo* landing on the Moon.

> *As to Mars, the possibility of life (and certainly of fossil life) should not be written off as quickly as some authors might admit. We have as yet sampled only two (very similar) habitats; there remain many enigmas of fossil drainage patterns, as well as the unexplored polar cap zones.*
>
> —*Joshua Lederberg, Nobel laureate, Rockefeller University, 1985*

Scientists would have been happy to find even microbes living on Mars. Any sign of life, though primitive, would be proof that life could arise on two different worlds, which would enormously strengthen the theory that life may be common in the universe. Indeed, *Viking* was designed precisely to answer the question raised by Percival Lowell in the 19th century, who thought the "canals" of Mars were built by a great civilization: Does Martian life exist?

On board *Viking* were a set of ultrasophisticated, miniaturized laboratories designed to look for signs of living organisms. Like any good robot, *Viking* had an arm, one rolled up like a carpenter's ruler. It slowly unwound until the claw at the end was able to dig a hole in the surface of Mars like a toy clamshell crane. The sample was then moved back into the lander and dumped into the *Viking* equivalent of a Cuisinart.® This processor distributed the sample to several different experiments.

One deposited Martian soil into what scientists fondly call chicken soup, a broth of chemical nutrients designed to encourage life to grow. Another applied radioactive carbon dioxide gas to see whether any life form took up the gas the way plant life would on Earth. Another experiment broke down the dirt into its chemical components and analyzed them.

Excitement reigned behind the scenes when two experiments actually seemed to detect life. However, eventually, most scientists concluded that these were false alarms caused by certain complex chemical reactions unexpectedly activated when the Martian soil came into contact with water. Because the Martian atmosphere is so thin, dangerous ultraviolet light from the Sun can reach the surface and cause exotic forms of chemicals to exist on the Martian dirt. These chemicals caused unexpected reactions in the experiments on board the *Viking*, and did not actually indicate life, although at least one scientist still argues that the reactions could well have been a sign of microbial life. However, not much carbon was found, which is the basis of all life on Earth. Most scientists agree today that the *Vikings* did not detect life on Mars.

Viking 2 landed in the region of Mars known as Utopia, an optimistic name given before the landing to a region that turned out to be noticeably different from the first site. This time *Viking* landed on the edge of a crater or boulder and was tilted, but fortunately not so dangerously as to threaten the experiment. The terrain was littered with pitted rocks looking very much like lava, possibly produced by some past volcanic eruption or a meteorite impact. *Viking 2* also found no signs of life in the samples that it tested.

Although most scientists think we can now rule out life on Mars, we should not forget that both *Viking* craft landed not at the best places to look for life, but at the best places for a safe landing. For two decades, NASA has sponsored designs of Mars *Rovers*, robotic explorers with wheels or tractor treads that would enable them to land at a safe place on Mars and then travel to more interesting sites.

It's an extremely difficult project, since Mars is so far away that it may take radio signals half an hour to get from there to here and back. If a *Rover* is about to roll over a cliff, that's disastrous, so it must have artificial intelligence smart enough to detect hazards and avoid them. So far, NASA hasn't convinced Congress to fund such a mission, but if the Russian and American Mars projects of the next few years are successful, we could see these *Rovers* taking off soon afterward.

Mars is covered with fascinating places. If we could explore the smooth riverbeds and ancient shorelines, we might find evidence— perhaps fossils of ancient creatures—that life once *did* exist there. Just that evidence in itself would almost prove that life must exist elsewhere in the universe, since conditions are different enough between Earth and Mars to resemble a range of conditions likely to exist on other planets around other stars.

Studying such ancient life forms from another planet would also tell us a tremendous amount about how life evolves and perhaps provide profound new clues to the origin of life on Earth. Since there is overwhelming evidence that water was once abundant on Mars, and since we know that life on Earth arose in the oceans, it is perfectly reasonable to suppose that there were once Martian fish, perhaps Martian octopi, and all kinds of wonderful and strange swimming things. On Earth, after swimming things had evolved, some of them started crawling out on land, and their descendants are the creatures that walk, run, and fly on our planet today. Were there ever Martian lizards or birds or dinosaurs? Or Martian dolphins or whales? A *Rover* might begin to answer these questions.

Another fascinating place to explore would be near the ice caps. We know that they contain frozen water, in addition to frozen carbon dioxide (dry ice). Perhaps on the edges of these ice caps some form of primitive microbe or moss or lichen lives today. *Viking* orbiter pictures show evidence that water probably exists underground on Mars, at least in the form of permafrost. It would be wonderful to be able to sample these regions to look for some form of life underground there.

The *Mars Sample Return Mission* is one of the most exciting on the horizon now in the space program. Scientists have proposed that the U.S. and the U.S.S.R. send a robot probe to Mars, where it would land and extract samples from the soil. A smaller rocket, carried on board, would then be launched with the sample and would fly back to Earth

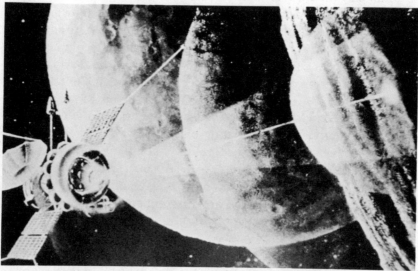

The Soviet Phobos spacecraft that died in 1989 while exploring Mars and Phobos. This artist's conception shows the laser beam shooting at the moon, vaporizing materials (represented by the two spherical waves centered on the bright spot on Phobos). The ejected particles would have been analyzed by instruments aboard the spacecraft. Some pictures of Phobos were sent before the failure.

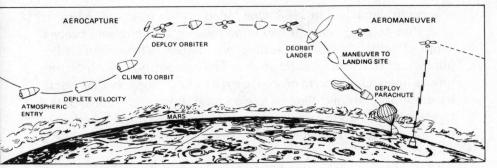

The proposed robotic *Mars Sample Return Mission.*

for recovery. It would be the first piece of another planet ever returned, and would allow us to test directly for microbes and microfossils. (Some meteorites on Earth are suspected to have come from Mars, but the shock of being knocked out of Mars and then hitting Earth would have destroyed any signs of life if they had existed.)

The *Mars Sample Return Mission* would combine the strengths of the U.S. and Soviet space programs. The Russian program puts up lots of basic hardware regularly, while the American space program is strong on high-tech know-how.

And although it is tempting for Americans to think the Russians have nothing to teach us, the fact is that they have accomplished some remarkably sophisticated missions that are as impressive as any we have done, such as their *VEGA* spacecraft that landed probes on Venus, put balloons in its atmosphere, and sent probes to Halley's Comet.

It's important to note, too, that the Russians are the only ones who have actual experience with rovers and sample-return missions to another world. Most of us have forgotten it, but the Russians landed three rockets on the Moon that took samples of the lunar surface and then returned to the Earth with them. On the *Luna 17* and *21* missions, their *Lunakhods* (lunar rovers) traveled as much as 23 miles on the Moon's surface. They returned a hundred thousand television pictures to Earth.

Despite the frequent surges of political hostility between the two countries, it is encouraging to know that there has recently been a large amount of scientific cooperation, sometimes encouraged by both governments.

Two Martian missions are officially approved now, one Soviet and one American. The Russians built a pair of very sophisticated spacecraft, launched in 1988. One failed on the way, and the other died after orbiting Mars. It had numerous experiments and probes, some of which were for landing on Phobos. They would have photographed its surface and analyzed its composition by vaporizing tiny samples. One bizarre probe it carried was a "hopper" that would have bounced around Phobos like a rabbit, taking advantage of the feeble gravity. This mission was the first chance to prove whether or not the moons are made of the carbon compounds astronomers suspect.

Then, NASA plans to launch the *Mars Observer* to study the geology and climate of the planet. Approved by Congress, it was scheduled to be launched in 1990, but on January 2, 1987, as part of a 15 percent cutback in planetary exploration, NASA announced it was postponing the spacecraft by two years, even though it doesn't need a Shuttle launch. The Planetary Society, among others, recommended that the project be restored to its previous schedule, and public and Congressional support for this mission was strong, but it still was postponed to 1992. Delaying *Mars Observer* will probably delay future exploration of the planet, including possible human flights.

The Soviets also plan other robotic Mars missions in 1992 or 1994, indicating their commitment to the red planet.

A fascinating new Russian-French-American design is being explored by Caltech scientist Bruce Murray, a participant in the Soviet Phobos mission. A novel "thermal" balloon would drift across Mars. Each day, solar heating would buoy it up into the atmosphere. At dusk, it would settle back down, taking pictures. This could be a low-cost way to explore Mars over vast distances—perhaps the next step in the robotic exploration of Mars. It will probably be launched by the Russians in the next several years.

The New Martians

The greatest human adventure likely to be undertaken in our lifetime is a trip to Mars. Many serious discussions have taken place between the Western spacefaring nations and the Soviet Union to explore the possibility of sending humans to Mars.

Way back in 1948, Wernher von Braun himself did a careful study of how to send an expedition to Mars, which he called *Das Marsprojekt*. He said then, "The logistic requirements for a large elaborate expedition to Mars are no greater than those for a minor military operation

As a focus for the development of new capability, we recommend the United States accept the long-range option or goal of manned planetary exploration with a manned Mars mission before the end of this century as the first target.

—Space Task Group (Vice President Spiro Agnew, chairman; Robert C. Seamans, Secretary of the Air Force; Thomas O. Paine, NASA Administrator; Lee A. Dubridge, Presidential Science Adviser), September, 1969

extending over a limited theater of war." However, in 1962, after he had gained many years of actual experience building more powerful rockets than he ever had in Germany, he modified that to, "I am now ready to retract from this statement by saying that on the basis of technological advancements available or in sight in the year 1962, a large expedition to Mars will be possible in fifteen or twenty years at a cost which will be only a minute fraction of our yearly national defense budget."

The main trouble with traveling to Mars is the distance. Mars lies about 50 million miles outside the Earth's orbit, about twice as far away as Venus. So Venus would actually be easier to get to than Mars, but it has a much nastier climate, and hence is not a place to which we'll send astronauts in the near future.

But even 50 million miles doesn't tell the whole story. This is 200 times farther away than the Moon, but that's only the distance we would have to travel if we simply shot from the Earth to Mars *when Mars was closest.* And that trip, unfortunately, would require far more rocket fuel than is currently conceivable to carry. We have to find a more economical approach.

The most straightforward way to get to Mars is to use a transfer orbit, a path in the form of a smooth ellipse that touches the orbits of Earth and Mars halfway around the Sun. This would increase the actual travel distance to about 400 million miles, making the travel time for a one-way trip to Mars about a year. Then you'd want to spend some time there, and would need another year to get back, so the round-trip time for a Mars expedition would be between two and three years.

Question: *You've covered our first 25 years in space as far as the public is concerned. What do you see for the next 25 years? If someone—one of your friends in Washington— were to ask you what thrust NASA should take over the next 25 years in planetary exploration, Earth observation, and communications, what do you see?*

Jules Bergman, television journalist: *I would say two things: Mars and the manned space station. We've got to settle the Mars thing. There has to be a companion planet to flee to when some nut starts World War III, because there'll be no place to hide. Well, let's hope that never happens. But that's one of the things I worry about.*

This is a long time, even for those of us who would love to go to Mars (and this author would be first in line to volunteer). You might point out, of course, that this investment of time is not so different from that once spent by sailors on worldwide journeys. However, remember that the astronauts would spend much of that time in zero-gravity, and then they'd have to endure the rigors of acceleration and deceleration, followed by walking around on a planet where they would weigh about one-third of what they do on Earth. Then, after they finished the exploration, they'd endure another year's journey back to Earth in zero-gravity.

One way out of this dilemma would be to spin the ship to simulate gravity, although that would complicate the design and make it harder to perform observations of the universe en route to Mars. Cosmonauts have endured almost a year in zero-gravity, so we know it is possible, with regular exercise, to stave off some of the weakening effects. Still, even the Russians' experience doesn't answer our questions about the stress of landing on Mars and conducting a full-fledged exploration.

The length of the trip also means that an enormous quantity of food and air must be brought along. (Three years is a lot longer than the one week it takes for a Moon trip!) Recycling would be of paramount importance. If we could grow food on board the spacecraft and convert our exhaled carbon dioxide back to oxygen (perhaps with the aid of plants), and if we could purify waste water to reuse it, then the amount of consumables would be drastically reduced. All we're asking the spaceship to do is essentially what the Earth does for us. We inhale oxygen, then exhale carbon dioxide, which in turn is inhaled by plants that convert it back into oxygen. Our waste water evaporates and eventually returns to us in the form of purified rain. And, of course, plants can be fertilized with waste, then eaten when they are grown.

All these supplies, however economical, would mean a great deal of extra weight, demanding yet more fuel than the already huge amount required to carry the astronauts and all their equipment. However, there are some ingenious tricks to cut down on the fuel needed. The atmosphere of Mars offers potential help that the vacuum of the Moon does not. First of all, it allows us to slow down without the use of much fuel, by braking in the Martian air, letting friction decelerate the spacecraft, just as the *Apollo* spacecraft slowed down on reentry by using the Earth's atmosphere, saving great quantities of fuel.

Second, the Martian atmosphere is mostly carbon dioxide. This has led to the ingenious idea that the Martian atmosphere itself could be used as fuel. The trick would be to break down the carbon dioxide into carbon monoxide and ordinary oxygen. This might be done with solar energy or a nuclear reactor. The oxygen could be used to renew the breathing supply on Mars, lowering the amount needed to be brought along.

But the most valuable fact here is that carbon monoxide will burn in oxygen; it's nothing but imperfectly burned carbon dioxide to begin with. (The carbon monoxide that comes out of your automobile exhaust is the result of imperfect burning inside the engine—fuel that wasn't completely oxidized.) This means that, in theory, carbon monoxide and oxygen could be used as rocket fuel. Designs have actually been made that would allow a rocket to burn carbon monoxide and oxygen. The astronauts could bring with them rockets designed to use this unusual fuel, and that's how they could take off from Mars and return to their orbiting mothership.

Orbital Alternatives

Scientists have also been investigating exotic types of orbits that could be used instead of the straightforward transfer orbit. One kind would involve a large vessel permanently orbiting between Earth and Mars, like a shuttle-bus that continually circles between downtown and the airport. The shuttle would be well stocked with supplies. The astronauts could dock with it when it was close to Earth and stay on board as if it were a flying hotel, then get off with a smaller spacecraft when they reached Mars.

A whole new class of orbits has been recently discovered that uses the Earth's gravitational field to boost spacecraft into orbits that could not otherwise be reached on limited fuel. At first glance, it would seem impossible to use Earth's gravity to get anywhere, because it simply drags you back to where you came from. However, Earth is moving around the Sun at almost 70,000 mph. Any spacecraft that leaves the Earth is thus already moving at that speed, plus whatever speed its rocket engines give it, minus the drag of our gravitational field. However, if you leave Earth and go out into space past the orbit of Mars and come back again, then—if you are clever enough and pick your orbit carefully—you can whip around the Earth into a new orbit that gives you some of our planet's speed, just as a Ping-Pong® ball hitting a paddle moves away with a combination of its initial speed plus the speed of the paddle.

The Earth becomes a giant Ping-Pong® paddle in this maneuver, which is known as Delta VEGA (meaning delta V—a change in velocity—with Earth Gravity Assist).

Then, too, Robert Farquhar of NASA's Goddard Space Flight Center in Maryland has suggested using the Moon's gravitation field for a similar, but smaller, effect, a trick requiring further study before it will be feasible to use on Mars missions.

He used this effect brilliantly in his design of the first spacecraft mission to fly by a comet—the project known as *ICE* (International Comet Explorer). He took a spacecraft that was orbited in the Earth-Moon system, never designed to leave its humble locale, and passed it by the Moon five times in a way that had never been done before. This gave it enough energy to leave the Earth-Moon system and fly on to meet Comet Giacobini-Zinner in 1985, half a year before the Russians flew by Halley's Comet.

One Step at a Time

Even though the first Mars explorers would undoubtedly like to land there, the first piloted trip may not land on the surface. It may be easier to rendezvous with the moons of Mars. There are two moons, Deimos and Phobos, tiny chunks of rock, ten to twenty miles

THE Ph.D. MISSION

The Martian satellites are of unique interest: as platforms for detailed Mars exploration and for later Mars habitation, as low-cost accessible sources of materials for propellants and construction, and as objects of intrinsic scientific value. Their origin is a mystery—whether captured or formed in place, perhaps remnants of an original larger body, or perhaps unmodified examples of primitive planetesimals.

A manned Ph.D. [Phobos/Deimos] mission can accomplish the scientific and resource objectives sooner and at lower cost than either a long series of unmanned missions directed from the Earth or a series of manned landings on the Martian surface. The Ph.D. mission can direct a series of unmanned rover vehicles to all parts of the surface and recover samples, including cores, for analysis in a Deimos-based laboratory. This sequential experimentation permits exploration of the most promising, and perhaps most difficult, terrain for a study of the geological, climatic, and possibly biologic history of the planet Mars. Data on past climates could shed light on climate processes on the Earth. It is widely believed that Mars was initially wet and warm, providing an environment suitable for the development of life forms. The discovery of fossil life or cryptolife would represent a major scientific achievement.

The Ph.D. mission promises to be no more difficult or costly than proposed manned missions to the Moon or to earth-crossing asteroids. To fit into an early timeframe, it would use chemical propulsion plus aerobraking, but would build on the experience of the U.S. and Soviet space stations to deal with the habitat and zero-g problems of a 2-year round trip.

—S. Fred Singer, George Mason University

in diameter—probably captured asteroids. The first astronauts to Mars will be likely to go to one of these moons and explore it, a fascinating task in itself, one that could give us great insights into the resources available to us in the solar system and also potentially valuable scientific knowledge about how the solar system formed. Such a mission

THE MILLENNIUM PROJECT: MARS 2000

No matter what other justifications may be given, the ultimate rationale for today's generations to return to deep space and to establish a permanent presence there is to create the technical and institutional basis for the settlement of Mars. . . . This expedition could be on its way by the end of the first decade of the third millennium. . . .

Why the hurry? Why a Millennium Project that stretches our reach to the limit? The answer is in the minds of young people who will carry us into the third millennium. It is in the generations now in school, now playing around our homes, now driving us to distraction as they struggle toward adulthood. They will settle the Moon and then Mars. They will do this because they want to do this. They want to "be there." Our role is merely one of staying out of their way while we preserve and expand their opportunities. . . .

There is little technical distance between us today and the realization of all that I have suggested. Certainly, there is little to be done compared to the task that faced us when we began the race to the Moon. Whatever technical options may turn out to be appropriate, now is the time to create those options so that the next generation may proceed when they are ready.

The first martian base will probably be established using inflatable shelters in one of the deep equatorial valleys near areas that show strong photographic evidence of being underlain by water-rich permafrost. The Valles Marineris is such a valley. The deep valleys also will provide somewhat higher temperatures and atmospheric pressure than other possible sites such as the polar regions where water-ice is clearly present. However, one of the advantages of using a large interplanetary space station for the first trip to Mars is that the time necessary to examine the martian moons for resources and to select a proper site for the first base can be spent in orbit about the planet before committing to the first landing. If necessary, reconnaissance of several sites can be carried out before committing to a site for the base. . . .

The confidence we can have in discussing the establishment of a permanent base on Mars comes from two direc-

tions: the confidence and knowledge gained from the Apollo expeditions to the Moon and the spectacular and detailed data returned by the Viking landers and orbiters of Mars. We know nearly as much about Mars as a planet as we did about the Moon before Armstrong and Aldrin landed there in 1969.

However, space activities will be sustained less by technology and knowledge than by emotions: the emotions of young Americans and young people the world over. Indeed, even young Soviets, East Europeans, Chinese, and Cubans also must look to space as the Earth's frontier. As with our ancestors, their freedom lies across a new ocean, the new ocean of space.

The Millennium Project: Mars 2000 is their hope as well as our mission.

—*Harrison Schmitt,* Apollo 17 *astronaut and former U.S. Senator*

would sidestep the entire complex question of how to land on, explore, and take off from the surface of Mars.

Then there's the easiest piloted Mars mission of all—a flyby of the planet. After all, our first mission to the Moon was a fly-around, *Apollo 8*, in which three astronauts orbited the Moon and then returned to the Earth. Such a mission would test the ability to endure the hardships of such a long trip, paving the way for a landing.

Some Western analysts suspect the Soviets may be planning such a Mars mission. It would explain why they have concentrated so much on prolonged orbiting of the Earth by cosmonauts. If they did it in the year 1992, they could celebrate the 75th anniversary of the Russian Revolution, and at the same time restore the face they lost when the U.S. beat them to a Moon landing. If they can't do it then, they might choose a later date, such as 1996, in which there will be a particularly good orbital configuration, or the year 2000, when the temptation will be to celebrate the new millennium with some spectacular achievement.

The possibility of a human expedition to Mars is so exciting that it has attracted great interest in the spacefaring nations. With advances in technology, the cost can probably be brought down to a figure less than the *Apollo* program, but it still would be extremely expensive.

LET'S GO TO MARS TOGETHER

. . . imagine a different sort of Apollo program, in which cooperation, not competition, was the objective, because the leaders of the U.S. and the U.S.S.R. had come to their senses. Imagine these leaders deciding to do something not just for their nations but also for their species, something that would capture the imaginations of people everywhere and would lay the groundwork for a major advance in human history—the eventual settlement of another planet.

It can be done. It is technologically feasible. It requires no major breakthroughs. A project to send people to Mars sounds absurdly expensive. But the advances in technology have been so great that such a mission would cost far less than Star Wars, less than the Apollo program, and not much more than a major strategic weapons system. In a joint mission, the cost to any one nation would be still smaller.

But why a joint mission to Mars? Why not jointly feed the hungry in sub-Sahara Africa, or do water reclamation projects in Bangladesh? The United States and the Soviet Union could, if they chose, together help house, educate, provide medical care for and make increasingly self-reliant every citizen of the planet. But the U.S. and the U.S.S.R. have no such precedent; they have been obsessed by the pursuit of short-term competitive advantages. The political realities, sadly, are that a joint mission to Mars, like Apollo/Soyuz, is well within the realm of practical possibility, while many worthy and more mundane activities are not. Not yet. But a major cooperative success in space can serve as an inspiration and spearhead for joint enterprises on Earth.

Moreover, space missions have an important subsidiary advantage: They use precisely the same aerospace, electronics, rocket and even nuclear technologies as does the nuclear arms race. . . .

In the long run, the binding up of the wounds on Earth and the exploration of Mars might go hand in hand, each activity aiding the other. The wonders of Mars will occupy us for a long time—its surface area is equal to the land area of Earth. The first voyage of men and women from our planet to Mars is the key step in transforming us into a multi-planet species—a step as momentous as the coloniza-

tion of the land by our amphibian ancestors some 500 million years ago and the descent from the trees by our primate ancestors perhaps 10 million years ago.

Decades ago, Mars called to the Soviet spaceflight pioneer Konstantin Tsiolkovsky and to his American counterpart, Robert H. Goddard. The rockets they designed were intended not for the destruction of life on Earth but to take us to the planets and the stars. Is there not some special obligation of the two principal spacefaring nations—the two nations that have burdened our planet with 55,000 nuclear weapons—to put things right, to use this technology for good and not evil, to blaze, on behalf of every human being, the trail to Mars and beyond?

—Carl Sagan, Cornell University

A joint mission would allow us to share the costs and the strengths of the different space programs. The time is ripe for the kind of East-West cooperation we've rarely seen since the *Apollo-Soyuz* project in 1975.

It's ironic that a period that witnessed the rebirth of the Cold War also saw, with official approval from both countries, American experiments on board the Russian spacecraft that flew by Halley's Comet. To some, in fact, the greatest attraction of the possible joint mission to Mars is not scientific but political. In fact, Senator Spark Matsunaga of Hawaii was so captivated by the idea that he wrote a whole book, *The Mars Project*, advocating this mission. The idea of the American and Soviet governments being committed to a joint human exploration of Mars, forcing cooperation stretching over a decade or two, is appealing as a way to plant the seed for benevolent communication that could replace the all-too-frequent hostility that seems to be the norm in American-Soviet relations. Such contact might help expose more Soviet citizens to Western points of view censored out of their press, and to strengthen *glasnost*; on the other side of the coin, we might learn to value more of their potential contributions to the world.

Mars is a planet packed with mysteries. Someday, probably in our lifetime, we will see men and women land there. The chances are that it will be a crew of Americans and Russians who first explore that world. From the comforts of Earth, billions of people will watch on their television sets as these explorers become the first human beings in history to walk the red sands of Mars, to probe the mysterious canyons, to climb the strange wedding-cake Martian ice cap.

It is only a matter of time before we establish permanent settlements there. We may start by setting up a permanent base, much as we have in Antarctica. With a whole planetful of mysteries, the ideal way to solve them would be to have people spend a year or two at a time there, the way scientists now do at Antarctic bases.

Once we have such a base, we will naturally learn more about how to live there, how to use the Martian resources of air and minerals and water to our advantage. Inevitably, children will be born—the first Martians. Some of these people will probably prefer to live there, instead of returning to an overcrowded, aging planet. Martian colonies will take root and spread until whole cities arise, making science fiction a living reality. Civilization will have a new planet and from then on, even the destruction of Earth would not obliterate humanity.

The red star with the magical name of Mars beckons to us irresistibly, filled with mystery and promise.

NINE

Dinosaur Killers?

ASTEROIDS AND COMETS

There is a place in space where vast resources are available even closer, in a sense, than the Moon. This place is among the asteroids—the great mountains of rocks that mainly orbit between Mars and Jupiter. Although physically located much farther away than the Moon, some are "closer" in terms of the amount of energy it would take to get there and back.

A few of them are in orbits that cross Earth's, making them relatively easy to get to. Some of them would be wonderful places to explore, as well as to harvest resources for the Earth and for potential colonies in space. The *Galileo* spacecraft will probably fly by an asteroid on its way to Jupiter, giving us the first close-up pictures ever taken of such a body.

Asteroids and comets are strange little mineral and ice bodies that litter the solar system. They seem to be left over from the formation of the system, chunks that were not swept up by the planets, and so are "fossils" of our beginnings.

Most asteroids flock near an orbit where a planet *ought* to be, in the wide gap between Mars and Jupiter. They seem to be the remains of a handful of Moon-sized planetoids that never grew into full-fledged planets, but instead smashed into each other, breaking into pieces up to hundreds of miles in diameter.

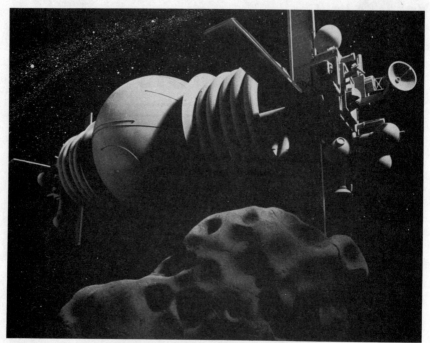

Artist Joel Hagen's picture of a space colony approaching an asteroid, where vast resources may be found.

Comets are dirty ice-balls of frozen water, ammonia, and methane (natural gas), a few miles in diameter, that formed out beyond the farthest planets, Pluto and Neptune. Temperatures are so cold there, near absolute zero, that they retain the most delicate compounds, probably including the chemical building blocks of life—ammonia, methane, water—that were here when the solar system formed. An unadulterated piece of a comet would be a clue not just to the formation of planets but to the very origin of life itself. It's the closest thing to a time machine for the study of our cosmic roots.

Pieces of comets and asteroids often strike the Earth, in fact, but are "cooked" by heat as they fly through the atmosphere and hit the ground. "Shooting stars" are usually pinhead-sized pieces of comet tails burning up in the atmosphere; meteorite craters are usually the result of a small asteroid hitting the planet. There is mounting evidence that our very oceans may have been largely formed by such bodies hitting the Earth billions of years ago.

The next time you drink a glass of water, pause and reflect on the likelihood that that liquid was once orbiting the Sun independent of the Earth. You are drinking space juice.

Comet Wars

During the last decade, there has been a friendly scientific battle among nations all over the Earth to be the first to visit a comet. NASA was never able to get its Halley's Comet mission funded, so the European Space Agency and Japan each built spacecraft of their own, and the Russians modified their Venus probe, *VEGA*, to do the same job.

Most comets actually stay out beyond Pluto in a huge cloud orbiting the solar system, called the Oort Cloud (named after the modern Dutch astronomer, Jan Oort). When the solar system formed, vapors of water and other substances condensed in the cooler parts of the system. The coolest place was out at the edges of the solar system, far from the Sun. So there the most delicate ices formed in snowballs, so to speak, orbiting around the Sun like billions of miniature planets. Most comets, the size of small asteroids, orbit the Sun quite slowly, taking millions of years to complete each pass. In the meantime, our Sun passes lazily through the Milky Way in its long trip around the center of our galaxy.

Occasionally, our solar system passes some other star, like two ships in the night. Although we don't get very close to one another, it can be close enough to disrupt the comets' orbits. They're so far from the Sun and the force of its gravity that they're susceptible to the feeble influence of another star's distant gravitational field.

Some of those comets may then be pushed into orbits that take them into the inner part of our solar system, among the planets. When these comets get closer to the Sun, they heat up. Ices that have lain in the frigid darkness of space undisturbed for billions of years begin to thaw and crack, to vaporize and explode.

Gas and dust start flowing off the comet, pushed back toward the coldness of space by the rays of the Sun and by the invisible solar wind of electrons and protons that streams constantly from the Sun. This is the tail we see when a comet passes across the sky.

In 1985–86, an international fleet of six spacecraft flew by two comets. The first was an American spacecraft not originally designed for comets, but ingeniously diverted, as described in the preceding chapter. Using the Moon's gravity, Robert Farquhar and his colleagues at NASA's Goddard Space Flight Center in Maryland pushed an old spacecraft, now called *ICE* (International Comet Explorer) into an orbit that took it past little-known Comet Giacobini-Zinner.

It was successful and made the first close-up measurements of cometary particle impacts, plasma electric currents, and magnetic fields. Unfortunately, it didn't have cameras, so the first close-up pictures of a comet came from the Russian *VEGA 1* probe of Halley's Comet. During the space of a week, five spacecraft flew past Halley: two Russian *VEGA*s, the Japanese *Suisei* and *Sakigake* probes, and the European Space Agency's *Giotto* probe.

They were all successful, giving us more information in a week than had been obtained in centuries of Earth-based observations. They found, for example, that Halley's Comet is one of the darkest objects in the whole solar system—as black as coal. It looks bright because we see only the glowing particles that have been shed by the dark nucleus. These spacecraft analyzed the cometary grains that hit them and discovered that the basic chemicals in some of the particles are the same ones out of which you and I are made: hydrogen, carbon, nitrogen, and oxygen. Comets that hit the Earth when it formed must have brought these chemicals with them, and probably helped—or even triggered—the formation of life.

As wonderful as these missions were, they gave us just snapshots of comets. They flew by these comets in the space of hours—cosmic one-night stands that told us about as much about the comet as a snapshot of a football game would tell an alien about that sport.

Another problem with such flybys is that because they examine the comet when it is active, they don't get a clear look at its nucleus. It's like trying to photograph a rabbit running through a snowstorm.

What we need to do is to rendezvous with a comet and stay with it for a long time. Ideally, we would start with a comet that is cold and quiet and inactive, and stay with it while it travels closer and closer to the Sun, heats up, starts to erupt, forms its tail, and does whatever magical things comets do that we have not yet seen.

The European Space Agency has recently made the decision to recommend a $400 million spacecraft to do just that, in a spectacular mission that would land a probe on a comet and bring back a piece to the Earth.

NASA's proposed *CRAF* (Comet Rendezvous/Asteroid Flyby) mission.

For several years, NASA has been studying low-cost missions, in order to see what space science could be done with the limited budget available for unmanned research. They have come up with a concept they call the *Mariner Mark II* series, and the first mission proposed is *CRAF* (Comet Rendezvous with Asteroid Flyby). *CRAF* would fly by and photograph asteroids on its way to rendezvous with a comet. A good candidate might be Kopff, a comet that orbits the Sun every six years.

The *Mariner Mark II* series will use modular spacecraft, or units developed for previous missions, plus selected new designs, to keep costs to a minimum. The spacecraft will be standardized, so that the same chassis design can be used for many different spacecraft by attaching different instruments.

CRAF, if approved, would be a very sophisticated mission, costing around $500 million—medium-priced, as robotic probes go. It would rendezvous with the comet while far from the Sun, and would put a probe down into it, and perhaps even land. It would then fly with the comet as it approached the Sun and take pictures of it as the tail started to form. We'd see for the first time the surface of a quiet comet, the black ices twisted into bizarre shapes by centuries of heating and cooling, and would watch the comet heat up, its surface cracking, geysers exploding into space.

Gold Mine in the Sky

Asteroids are easier to get to than comets, since some have orbits not very far from Earth's. Most likely, humans will visit asteroids before they get to comets.

Asteroids range in diameter from under a mile—just barely detectable in the most powerful telescopes—to 600 miles, the size of Ceres, the first asteroid discovered.

Since asteroids have almost no gravity, it would be relatively easy for a spacecraft to rendezvous with one. Most are so small that if you jumped off the surface, you would escape forever from the object.

More than 2,000 of these flying mountains have been located, and undoubtedly many thousands more exist. Some are made largely of rock, others of a surprisingly pure nickel-iron alloy, and others of carbon compounds somewhat like tar. Practically speaking, the metals and the carbon compounds are potentially valuable resources; and as for scientific benefits, asteroids must contain clues to the mystery of their origin and the formation of planets. Perhaps carbon compounds carried by asteroids contributed to the formation of life here.

These carbonaceous asteroids may well be one of the most useful resources in the solar system. In theory, they could be used to make oil, synthetics, even food, and water might be extracted from them. So a carbonaceous asteroid could provide many of the important resources for a space colony; nickel-iron ones might supply the building materials.

Because the gravity of an asteroid is so weak, we could land on it almost effortlessly and take off again with a minimum of fuel. That gives the asteroid an advantage over the Moon, where you have to fight gravity to keep from crashing into it, and then you have to fight gravity again to take off.

Since the *Apollo* landings on the Moon, a number of scientists have suggested that we ought to land astronauts on an asteroid. This would have enormous scientific value, as we would then be able to examine an asteroid in the raw for the first time, instead of having to be content with the tiny fragments that pass through the Earth's atmosphere to become meteorites cooked by our atmosphere's friction. Such a mission would also have potential value for the future of space colonization, because we could find out whether mining asteroids would be practical.

The Threat of Collisions

Asteroids also present a potential hazard to Earth. For example, one called Hermes passed only half a million miles from the Earth in 1937. That's just twice as far away as the Moon, incredibly close by astronomical standards.

Even the impact of a small asteroid could be disastrous. Many scientists now think, for example, that it was an object only about ten miles in diameter that hit the Earth 65 million years ago and created an enormous explosion, covering the Earth with a dark cloud that prevented sunlight from penetrating, drastically lowering the temperature, killing off much plant life, and extinguishing many animal species, including the dinosaurs. Such a collision would release more destructive energy than all the nuclear weapons on Earth today.

The smaller these objects are, the more likely one of them is to collide with the Earth, because they are so numerous. But the problem is that small objects are extremely difficult to spot from afar. Usually they cannot be detected if they are very small until they are relatively close to the Earth, at which time we would have little chance to react to them. Some scientists have suggested that we create a sky survey to watch for such asteroids, so we can prevent them from hitting the Earth. It is inevitable that sooner or later a small asteroid will hit us with the energy of an H-bomb; but if we could detect its approach in time, we could send missiles into space to divert or destroy it.

There may be other asteroid belts even deeper in space. A strange asteroid called Chiron was discovered in 1977 by astronomer Charles Kowal. This is the first and only asteroid found whose orbit lies between Saturn and Uranus, but it may be the brightest asteroid of an unknown belt out there.

NASA has already built a spacecraft that will probably fly by an asteroid. Known as *Galileo*, the spacecraft was supposed to be launched by the Space Shuttle in 1986 but was postponed following the Challenger explosion. *Galileo* is now sitting on the ground waiting to fly, as NASA studies various options. The main purpose of *Galileo* is

Picture an onrushing comet or asteroid that approaches
Earth at a relative velocity between 10 and 50 kilometers
per second, so that the object arrives from the moon's
distance in a matter of a few hours—not much of a warning
even if anything were to done about it. The comet or
asteroid would part the atmosphere like a superprojectile,
leaving a hole through the air as wide as itself. Even though
air rushes back into the hole at the speed of sound, it would
taken tens of seconds for the hole to disappear. Upon strik-
ing the earth, the object would decelerate to zero velocity
only after depositing most of its kinetic energy in the material that it encounters. Thus, the object would stop only after
it has encountered a total mass of material several times its
own mass—a bit more than this if we describe a high-
velocity comet, a bit less if we consider a relatively low-
velocity asteroid impact.

The oceans have an average depth of 7 kilometers, so the
ultimate result would be much the same whether the object
struck land or water; if the latter, it would plow a hole
through the water—after ripping through the atmosphere—
and then excavate a crater several kilometers deep and 50
to 100 kilometers wide in earth's crust. The water around
the comet would be vaporized at once, doubling the water-
vapor content of the atmosphere. Applying our knowledge
of earthquakes to the object's kinetic energy, we find that
the earthquake from such an impact would release 100
billion times more energy than did the 1906 San Francisco
earthquake. Another effect of an impact in the oceans
would be a tsunami (popularly misnamed a "tidal wave")
about a kilometer high, which would roll across the seas at
1,000 kilometers per hour and cause widespread destruction
in areas within a hundred kilometers or so of the coast.

Those are but the side effects; most of the damage to life
on Earth would arise from dust. A 10-kilometer object
would move about 200 cubic kilometers of rock, more than
a thousand times the amount excavated over ten years to
make the Panama Canal. The impact would heat this
material instantaneously, spraying it sideways and upward
from its crater. Matter heading upward would encounter no
resistance from either ocean (if the comet struck there) or
atmosphere, for they would have been thrust aside suffi-

*ciently for the material to rise unimpeded. The pulverized
particles would each acquire a ballistic trajectory, like the
path of a rocket once it has left the atmosphere and its
engines have shut down. Each particle rising high above the
earth would orbit our planet. Some of the particles would
fly off into interplanetary space, but most of them, still held
by Earth's gravity, would fall back onto the top of the atmosphere, at a point far from the hole through which they
emerged. The heavier particles would fall through the atmosphere at points all around the globe, but the lighter ones
would float like oil on water, suspended in the stratosphere
as fine-grained, low-mass dust grains. The result would be
death for many species of life on Earth.*

—*Donald Goldsmith*, Nemesis

to go to Jupiter, but on its way there, it must pass through the asteroid belt. Originally, the plan was to fly by one or two particular asteroids and take the first close-up pictures ever made of those bodies. This will still be done when *Galileo* is finally launched, though NASA has chosen a different asteroid.

And the Russian Mars probe that was to vaporize bits of the moon Phobos was, in a sense, also an asteroid mission, since the two moons of Mars are probably asteroids that were captured when they wandered too close to the Martian gravity.

One day we may find it easier, cheaper, and less environmentally damaging to the Earth to mine metals and other necessities in space from asteroids or the Moon and ship them back to the Earth. There would be no need to strip-mine the Earth if we learned to use all the resources waiting out there, like goods in a supermarket, waiting to be taken off the shelf.

Some of the atoms in our blood probably came from asteroids and comets. The visits to these bodies, which will be taking place in the next few years, will be major steps in tracing our cosmic roots, unraveling our planetary genealogy.

TEN

Near Neighbors

THE INNER SOLAR SYSTEM

Mercury and Venus are the only two planets closer to the Sun than Earth. Both have been visited by space probes, and both have given us major clues about the history of the solar system, generating great debates about conflicting theories.

Recently, interest in Mercury has been growing, leading some scientists to propose sending probes there. Venus is the target for two spacecraft: *Magellan*, a NASA radar mapper now being built, and a possible Russian probe that would float a large French balloon in its atmosphere.

And then there is the Sun itself, central to our life yet disturbingly violating our theories. It will be probed by a West European spacecraft, *Ulysses*, showing us the poles of our nearest star for the first time.

A Hell of a Planet: Venus

The closest planet to Earth is Venus, the beautiful, bright world identified since ancient times with the goddess of love. Venus astonished the great 17th-century Italian astronomer Galileo by showing phases like those of the Moon: at times a crescent, at other times a full circle. To his powerful mind, this was proof that Venus was a world orbiting around the Sun and not, as his contemporaries claimed, orbiting around the Earth. The Polish astronomer Copernicus was right. Earth was not the center of the universe. But the Church disagreed, and Galileo suffered arrest and imprisonment.

Venus has mystified and fascinated astronomers ever since. Markings like those of Mars were allegedly observed by a couple of early astronomers, including one who, in 1727, claimed to have mapped the seas and continents on the surface of Venus. What he didn't know was that you cannot see the surface of Venus from Earth because of its thick, constant layer of clouds.

The first person to realize Venus had an atmosphere was a Russian astronomer, Mikhail Lomonosov, in 1761. He, no doubt, would have been delighted to learn that the first pictures ever taken from the surface of another world were those from a Russian spacecraft on Venus.

The thick, cloudy atmosphere of Venus has almost a hundred times greater pressure than the Earth's, and the constant cloud cover keeps us from seeing the surface. As a result, for centuries astronomers, trying to measure the spin of the planet, got results that ranged from about one Earth-day to a figure such that the planet's spin-period was equal to its year. It wasn't until 1961 that a correct measurement was made by radar. It turns out that Venus is spinning slowly backward.

Venus takes 225 days to go around the Sun, but it rotates so slowly that it takes 243 days to complete one period of rotation. Astronomers found in the 1960's that, despite the very slow rotation of this planet, there are clouds in its atmosphere that roar around the planet every four Earth-days, indicating very strong winds in the upper atmosphere, a puzzle to meteorologists who hope to improve their understanding of Earth's weather by studying the climate of other planets.

The race to explore Venus has been a hotly contested one. The first probe sent to Venus was Russian, but it died long before it got there. The first probe to succeed in flying by the planet was the American *Mariner 2*, and then the first successful lander was the Russian *Venera 7*.

The Russians have been as successful with Venus as the Americans have with Mars, persistently improving their spacecraft and accomplishing more with each mission. Thus, by 1982, the Soviet *Veneras 13* and *14* landed on Venus, and one of them lasted for more than two hours—a record for survival in the hot, nightmarish world of Venus. The temperature was around 900 degrees Fahrenheit. They found a bright orange sky and rocks of reddish brown.

The Mariner spacecraft now on its way to Venus is the most intricate instrument in the history of space science. The accuracy of that shot is comparable to firing a missile from Cape Canaveral and dropping it in [a] stadium between the 40-yard lines.

—*President John F. Kennedy, September 12, 1962*

Probably the greatest mystery about Venus is that it is so similar to the Earth, yet so radically different. Being about the same diameter and mass as the Earth, it was sometimes called Earth's twin in the past. So why does it have such a thick atmosphere? Why is it dry? Where are the oceans?

In fact, this series of questions can be extended to include Mars. Mars is just a little smaller than Earth or Venus, yet it, too, is radically different from either, with a much thinner atmosphere. Earth sits sandwiched between these two similar worlds. On the one hand, Venus has about a hundred times our atmospheric pressure, while Mars has about a hundredth. Both Mars and Venus have carbon dioxide atmospheres, yet the Earth's atmosphere is mostly nitrogen with, of course, a substantial amount of oxygen. Why these differences? Until we understand both neighboring planets, we will not entirely understand Earth.

Because of the spacecraft that have flown by and landed on these two planets, we understand some of the differences. Some scientists now suspect that the Earth's original atmosphere resembled the current atmosphere of Venus. Our ancient atmosphere may, for instance, have contained a lot of carbon dioxide, just as Venus does. Indeed, we now know that if all the carbon dioxide in the rocks of the Earth were released, we would be breathing an atmosphere very similar to that on Venus today, about a hundred times thicker than the air we now breathe.

Part of the answer seems to be that on Earth, the chemical formation of rocks used up most of our carbon dioxide, taking it out of the atmosphere and trapping it in mineral form.

We would like to pierce through those thick Venusian clouds, and

The approved NASA *Magellan* spacecraft, which will orbit Venus and provide picture-quality radar images of the planet's surface.

there is a way to do this without landing: radar. American and Soviet spacecraft have already made radar maps of the planet, but we would like to improve the quality of these images to the point where we can get images from the surface as clear as photographs. Thus, NASA has designed a spacecraft called *Magellan* to "look" through the atmosphere and take those radar pictures. This project has been approved by Congress and should be orbiting the planet in the early 1990's.

The design of the Venus radar was inspired by the Earth-orbiting satellite SEASAT. SEASAT orbited the Earth in 1978 and provided precise, detailed pictures of the Earth's surface and oceans. It transmitted a repeated signal while the spacecraft orbited, then added up all the reflected responses. It simulated an enormous antenna. (The amount of detail that can be obtained with radar is normally directly proportional to the size of the antenna, so a small antenna gives only a fuzzy picture. However, by this technique, called Synthetic Aperture Radar, engineers were able to make a small antenna act like a giant one.)

The Russians then used the same technique on one of their Venus missions, obtaining pictures so impressive that NASA was forced to upgrade the quality of the *Magellan* spacecraft so it wouldn't be a mere rerun of the Russian mission. *Magellan* will also use the SEASAT technique, covering almost the entire planet and obtaining radar images as sharp as photographs; it will be able to see details less than a mile in size.

We know there are Venusian continents—huge areas thrusting out of the lowlands—even if there is no water. Are they drifting on a molten interior like Earth's continents? What effects does this thick, forbidding atmosphere have on the development of mountains and valleys over billions of years? Venus is a world so similar to Earth, yet so different; surely many fascinating peculiarities will be brought out when *Magellan* flies around this planet.

The Russians had great success with their *VEGA* missions, dropping balloons into the Venusian atmosphere as they flew on to meet Halley's Comet. The balloons, designed by the French, had to be reduced in size to allow for the comet mission, but the two nations are considering building another Venus probe to put a much bigger balloon there that would float around the planet, making observations.

Venus is a world that we may one day transform into a place where humans can live. The atmosphere contains all the oxygen we would ever need, if we could just separate it from the carbon dioxide.

Battered World: Mercury

The closest planet to the Sun is the hot, dry world of Mercury. Located 37 million miles from the Sun and 40 percent of the distance between the Earth and our star, it speeds around the Sun in 88 days; it has been visited by only one spacecraft, NASA's *Mariner Venus Mercury*.

In the 19th century, Percival Lowell saw canals not only on Mars but on Mercury as well. However, since no other astronomer saw them, we may presume that Lowell, although basically a good astronomer, was a bit canal-happy: *Mariner Venus Mercury* proved there are no signs of an alien civilization on that planet.

The spacecraft, originally called *Mariner 10*, was designed to fly by Venus and then Mercury, which it did successfully. After it had been launched, scientists discovered that its orbit had a peculiar property

*If you ask, "Why do human beings explore?" I would
answer, as I think the Greeks would answer, "Because it is
our nature." . . . A clear context in which this was put for
me is a beautiful ethnographic work by a woman called
Edith Marshall Thomas, who lived for many seasons among
a small group of the wandering peoples of the Kalahari
whom we call Bushmen, people whose inventory of physical goods is very small indeed. They own nothing that sits
still. They carry all that they have, all that they make, in a
pouch of hide which they bear on their shoulders. They
wander forever through life, stopping now here, now there,
to sleep in a kind of nest, to try the fruit of this tree, to
scratch up that waterhole, to meet for a ritual encounter
with their wandering friends, and so on. These people,
whose minds are full, though absent writing, absent
crowds—in fact they are few—live in small bands of extended families. Each band tends to stay within a region about
like that of Los Angeles County, an area of a thousand
square miles or two, in quite desert country. From their
point of view they are by no means poor; they manage to
make an excellent living, as the time-and-motion study people have demonstrated to us, while working rather less hard
than the Harvard anthropologists who watched them. Their
skill is so great, their understanding and their wants are so
well controlled in the environment, they are so beautifully
adapted to their situation, that they need not work harder.*

*The one need they constantly discuss as they wander
through the cool mornings, the cool evenings, and as they
rest in the heat of the day, is to know exactly where they
are. They discuss it always. They note every tree, they
describe every rock. They recognize every feature of the
ground. They ask how it has changed, or how far it has
been constant? What story do you know about this place?
They recall what grandfather once said about it. They conjecture, and they elaborate; their minds are filled; their
speech elaborates exactly where they are. You see they have
built an intensely detailed, brilliant, forever reinvigorated
internal model of the shifting natural world in which they
find their being. What that simplified case suggests I dare to
extrapolate to all human beings everywhere. I see in it, I
think, my own behavior; I hope it will be so for others. It is*

> *fair to say that our language, our myth and ritual, our tools,*
> *our science, indeed our art, are all expressions translated in*
> *one way or another by the symbols of our communication*
> *or otherwise of certain features of this grand internal*
> *model. The presence of that internal model and its steady*
> *need for completion, the obviously adaptive need of its*
> *leading edges to have continuity, not to fade off into the*
> *nothing or the`nowhere: this is the essential feature of*
> *human exploration, its root cause deep in our minds and in*
> *our cultures.*
>
> *For me, exploration is filling in the blank margins of that*
> *inner model, that no human can escape making.*
>
> *—Philip Morrison, MIT*

causing the spacecraft to return periodically to Mercury. That is, its orbit was such that after flying by Mercury and traveling by the Sun, it returned to a position close to the planet.

As a result, NASA was able to fly past the planet three times. After the third pass, the spacecraft died and, no doubt, continues to make its periodic rendezvous with the planet Mercury, a modern flying Dutchman for the foreseeable future.

The first thing that became obvious about Mercury was that it was a nasty place. Not only was it terribly hot, but it was a vacuum. There was virtually no atmosphere on the planet. The photographs sent back by the spacecraft showed it to be heavily cratered. Indeed, it was hard for anyone but an expert to tell a picture of Mercury from one of the Moon. The temperature ranges from 660 degrees on the day side to 270 degrees below zero at night (Fahrenheit).

But those craters are important. On the surface of Mercury, meteorites have written a record of the formation of the solar system, and by comparing them with the craters on the Moon, Mars, and other bodies, we begin to see differences in the conditions that existed when the planets formed.

Originally, there was nothing but dust and gas in our part of the galaxy. Then, a cloud of dust began to collapse, drawn into itself by the gravitational attraction of the matter for itself. It compressed and became a huge ball of gas and dust that heated up as the particles rammed into each other, competing for the gravitational center. Eventually, the particles heated up so much that they reached a point at which hydrogen gas began to fuse as it does in a hydrogen bomb, releasing enormous energy.

Debris continued to orbit around this forming Sun and the debris itself was drawn together into various masses, forming balls of gas and dust that collapsed into planets such as Mercury and Earth.

Much of the debris was in the form of rocks that flew throughout the solar system like a great game of billiards, smashing into one another and hitting the planets and the moons, leaving the record of their passing like hieroglyphs carved on a pyramid. On the Earth, the record has been largely obliterated by the atmosphere, weathering, and the movement of the continents.

Mercury presents several mysteries. Why does it spin so slowly? Up until 1962, it was thought that Mercury was frozen in its rotation so that it would always keep the same face toward the Sun, just as the Moon keeps the same face toward the Earth. But measurements showed that instead of having an incredibly hot surface facing the Sun and a fantastically cold surface facing toward space—the subject of many science fiction stories—the dark side was warmer than it would be if that were the case. This suggested that Mercury was actually rotating in some way and was not locked into its orbit around the Sun. And then, in 1965, scientists at the Arecibo radiotelescope in Puerto Rico bounced radar signals off Mercury and measured the spin of the planet. They found, indeed, that it was slowly rotating.

Arecibo discovered that Mercury's rotation is also locked somewhat like the Moon's, but in a bizarre way. It's locked so that if one crater faces the Sun now, after one period of revolution around the Sun, that same crater will be facing directly toward space, toward the night sky, away from the Sun. Then, after one more period around the Sun, it comes back so that the same crater is now facing toward the Sun.

Why does Mercury spin in this strange fashion? If we look at the Moon, which is locked into a standstill, we find that it was slowed down over time by friction between the Earth and Moon, caused by the tides, through gravity. A long time ago, the Moon spun rapidly; the Earth undoubtedly spun much faster, too.

In fact, it's quite likely that all the planets originally spun with something like a ten-hour period. A day-night cycle back then would have been about ten hours long on any planet. Various forces, particularly tides, acted to slow down many of the planets. Some, such as Jupiter and Saturn, still spin in approximately a ten-hour period.

Earth has slowed down to its 24-hour day because its oceans are tugged on by our large Moon. That force slowed the Earth's rotation like a toy top on a rough table. At the same time, Earth slowed down the Moon's spin until it was locked so that it always kept one face to us.

Apparently Mercury was slowing down over billions of years in the solar system, but it hasn't yet slowed to the point where it is locked into a standstill. Instead, it's locked at an earlier stage. We suspect that the tidal force created by the Sun on Mercury inexorably slowed the planet's spin. We need a spacecraft to make more precise measurements of Mercury's gravity to test the theory.

We also know that Mercury has a weak magnetic field. This finding surprised many scientists because magnetic fields, we think, are caused by electric currents generated in the conductive spinning liquid cores of planets. Mercury spun so slowly that many scientists didn't expect to find a magnetic field there. True, it was, in comparison, a small field, about 1 percent of the Earth's magnetic field. However, this is stronger than the magnetic field of Mars, Venus, or the Moon.

The magnetic field tells us a lot about the currents of electricity and fluid in the interior of Mercury, and studying it helps us understand the Earth's magnetic field. Without other planets for comparison, we couldn't thoroughly test our theories of the Earth's interior.

Mercury has also recently been found to have a very thin, exotic atmosphere of sodium and potassium gas, perhaps knocked out of its surface by solar energy. NASA scientists are now devising ingenious orbits that could allow an inexpensive spacecraft with little fuel to orbit the planet, after bouncing off the gravitational fields of Venus and Mercury. Small probes could be launched to penetrate the surface to show whether there are Mercuryquakes there, which would tell us a lot about its interior—for one thing, how much of it is liquid.

The Star of Our Story: The Sun

The biggest actor in our history is the one we, in some ways, know the least about. For a long time, we thought that we understood the Sun, since we had an excellent theory of how it generated its energy from hydrogen.

Starprobe, a proposed NASA spacecraft that would travel to within 1.3 million miles of the Sun.

In the last few years, however, we've found out that something is seriously wrong with our theory. It's one of the biggest mysteries in all of science—as big as the 19th-century puzzle, when scientists couldn't figure out where the Sun got its energy.

We thought that Einstein had solved the mystery of the Sun for us. His equation $E = mc^2$ showed that if two hydrogen nuclei (protons) stuck together, they would release a great deal of energy. This is the principle of the H-bomb—thermonuclear fusion—and it seemed to be the principle behind the great energy of the Sun and most other stars.

But a scientist named Raymond Davis, Jr., of the University of Pennsylvania, built a "telescope" deep inside the Homestake Mine in South Dakota. Using 100,000 gallons of a chlorine compound, he searched for the fiendishly difficult-to-detect particle known as the neutrino, which should have been produced by the thermonuclear reactions in the Sun. Once in a great while, a neutrino from the Sun hits a chlorine nucleus, converting it into argon, a gas detectable by its radioactivity. He should have been able to detect the Sun's neutrinos.

But he didn't.

Other scientists have built similar neutrino telescopes and have finally detected some solar neutrinos, but only one-third the number expected. What happened to the other two-thirds?

We don't know.

Physicists have tried to fudge the theories to produce smaller amounts of neutrinos, but nothing works. Either there's something wrong with our theory of the Sun, or neutrinos are not stable and transform into other particles (perhaps a type of undetectable neutrino) before they reach Earth.

There is no body in heaven as important to us as the Sun. Even a tiny change in its operation could cause an ice age or could melt the ice caps, flooding the coasts of the world. Perhaps such events have governed the radical climate changes in the past.

It's very difficult to get close to the Sun to study it. Electronic systems break down when they get very hot, which is why the Russian Venus landers functioned only for an hour or two on the surface of that planet. The West Germans built two spacecraft called *Helios* that got closer to the Sun than any others before them. Launched by NASA in 1974 and 1976, they got to within 30 million miles of the Sun—closer than the planet Mercury—and returned data successfully.

But we'd like to get even closer, and ideally, fly over the poles of the Sun, which we can't see clearly from Earth because of the hot, turbulent solar atmosphere. We have evidence from the behavior of the solar wind that the Sun behaves differently at higher latitudes, so it is particularly important to fly over the Sun and look down, to see what's going on in its "arctic" (though still incredibly hot) regions.

A few years ago, the U.S. and the European Space Agency proposed to build two spacecraft to do just that. It takes a lot of energy to propel a spacecraft out of the plane of the planets' orbits and over the solar poles, and the scientists came up with the seemingly crazy idea of shooting the spacecraft *away* from the Sun. By going to Jupiter, they found, they could use its immense gravitational field to swing the craft into orbits that would take them out of the planetary plane and over the Sun's poles.

After the two agencies started work, the U.S. budget was slashed, killing the American spacecraft and annoying our European friends no end. They continued, however, and built one of the spacecraft with

U.S. help. It was to be launched by NASA in 1986, until the *Challenger* disaster forced postponement.

The spacecraft, called *Ulysses*, is now scheduled to be launched. In 1990, it will begin flying toward Jupiter on the first leg of its journey over the mysterious poles of the Sun.

On the back burner, NASA has designed on paper a *Starprobe* that would approach within a hellish million miles of the Sun. This would be the logical follow-up for *Ulysses*, and could tell us about the distribution of matter inside the Sun, by precise measurement of its gravity. *Starprobe* could provide crucial clues to the most important object in our universe.

ELEVEN

Giant Worlds

THE OUTER PLANETS

Beyond the asteroid belts lie the giants of the solar system. Out there, in the cold, dark recesses of the solar system where the Sun is small and distant and the planets take decades to orbit, are the four giant worlds of Jupiter, Saturn, Uranus, and Neptune, along with tiny world Pluto. Jupiter, Saturn, and Uranus have all been visited by spacecraft, and Neptune will be reached by the *Voyager 2* probe in 1989. Jupiter will be revisited in the next few years by the *Galileo* orbiter, which will send a separate probe to plunge into the atmosphere of the planet. Saturn may be revisited by a similar probe, but one that would enter the atmosphere not of the planet but of its unique moon, Titan.

We have learned an immense amount about these planets in our first reconnaissance of the solar system, and as usual in science, have uncovered as many mysteries as we have solved.

The Red-Eye Special: Jupiter

Jupiter, the closest of the giant planets, was first reached by the *Pioneer 10* spacecraft in 1973. That expedition was soon followed by *Pioneer 11* and, later, the two *Voyagers*.

The two *Pioneers* lived up to their name, returning the first close-up pictures of the planet and fuzzy images of its four large moons. They also sent back pictures of the enormous Great Red Spot on

Jupiter—a massive storm that's raged for centuries, so large that three Earths would fit inside it. It is one of the wonders of the solar system: What keeps it going? Is there something underneath that triggers it?

The four giant planets are mostly gas and liquid. What we see are the dense clouds on the outside, suspended above vast oceans of hydrogen and helium, ammonia, and methane. These are the same gases most scientists believe were present on Earth when life began to form.

Jupiter's magnetic field proved even greater than expected, extending millions of miles from the planet, a region larger than the Sun. Whirling around every ten hours, Jupiter measures eleven times the diameter of the Earth. This rapidly whirling magnetic field forms a strange flying-saucer-shaped magnetosphere, a region controlled by Jupiter's magnetic field, where electrons and protons are accelerated to energies measured in millions of volts, producing Van Allen radiation belts vastly more intense and energetic than Earth's, and forming a serious hazard for spacecraft. The *Galileo* orbiter contains instruments to measure these particles and to map this unique magnetosphere.

When the *Voyagers* flew by Jupiter, they took exquisitely detailed pictures of Jupiter and its moons. Jupiter's atmosphere turned out to exhibit a variety of strange phenomena. In addition to the still puzzling Great Red Spot, there were "daisy chains" of storms circling the planet, belts of clouds that had been seen fuzzily from the Earth for centuries, and immense rafts of clouds the size of entire planets, indicating some strange atmospheric phenomena.

The *Voyager* pictures of the large moons of Jupiter were real eye-openers. The fuzzy worlds photographed by the *Pioneers* suddenly came into focus, as clear as our own Moon seen through binoculars. Each of them turned out to be a unique world in its own right.

Each one is a high priority for exploration by the *Galileo* mission, which will pass far closer than the *Voyagers* did, revealing never-before-seen details of these moons' puzzling features. *Galileo's* ultraviolet instrument will also detect any planet-encircling gas-rings produced by its moons, telling us much about the interactions between Jupiter's satellites and its magnetosphere.

The outermost of the four moons, Callisto, is a ball of ice covered from pole to pole with crater after crater after crater. Its surface is pitted so much that a meteorite hitting this moon would have to obliterate some earlier crater to leave a new one in its place. Almost on the equator is an enormous bull's-eye crater called Valhalla, named after the great hall honoring immortal warriors in Norse mythology. About half the three-thousand mile diameter of Callisto, this huge bull's-eye commemorates the collision of some major piece of cosmic debris with this great moon.

Next in from Callisto is Ganymede, another icy world, but one that is covered with glacial highways. A strange network of features looking very much like a 19th-century map of the Martian canals covers the face of this world. Curious grooves line many of these highways of ice. Perhaps there was once a drifting of continents on Ganymede as there is on Earth—we don't know.

Next is Europa, looking like a glass eye just shattered by a cosmic hammer. The surface is completely covered with thousands upon thousands of long canal-like cracks. Yet it's comparatively as smooth as a billiard-ball, its cracks filled in with ice. But our most remarkable discovery about Europa (based on a combination of *Voyager* observa-

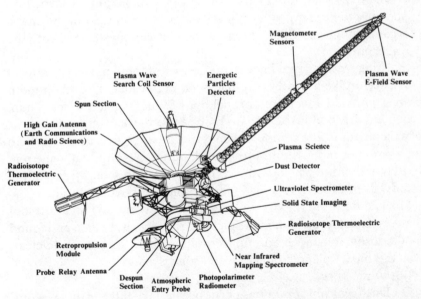

The *Galileo* spacecraft, already built, that will be launched soon to orbit Jupiter. The conical object on the bottom will separate and drop into the planet's atmosphere, returning the first measurements ever taken from within Jupiter's clouds.

tions and theoretical models of the interior) is that it probably has a liquid ocean beneath its surface. A water ocean was the last thing that most scientists expected to find on the icy moons of the outer planets, but this seems to be the case.

The explanation comes from the innermost large moon, Io.

===== Alien Volcanoes and Ice-Covered Oceans =====

Io is the most spectacular find of all, and will be a special target of *Galileo.* Before spacecraft visited it, there were hints to Earth-based astronomers that something about it was strange. Its color, for one thing, is orange, very unlike the grayish and brownish colors of most of the solid bodies of the solar system. Some astronomers reported that Io became suddenly brighter after it emerged from Jupiter's shadow, as if snow had fallen in the dark, cold night of the eclipse.

The instruments of the *Pioneers* revealed a ring of gas orbiting Jupiter like an invisible Saturn ring, clearly produced by Io. This was a clue that somehow that moon was leaking gas, which would escape into space only to be trapped by Jupiter's gravitational field in the shape of a giant doughnut. But its origin was to remain mysterious until the *Voyager* mission.

Startling pictures sent back from the first of these missions showed a surface unlike any other we had seen in the solar system. Smooth planes with odd, black craters pockmarked the orange surface. Then, when Linda Morabito of JPL was doing a routine navigational analysis of a picture of Io, the computer kept yielding strange results, as if there were another moon behind Io—but there wasn't. Gradually she realized that there was a "bump" on the picture of Io, a bump extending hundreds of miles into space—far too high to be a mountain.

She had detected the first volcano ever seen erupting on a celestial body. Scientists reexamined the pictures they had just taken and found half a dozen volcanoes erupting simultaneously. Io, it became clear, was the most volcanic object in the solar system.

But why?

On Earth, volcanic eruptions occur because the interior of our world is heated by radioactivity that keeps our planet molten inside. Occasionally, pressure from within forces that magma out through the surface. Io, however, was too small for this to be an important effect. Yet just before the first *Voyager* arrived at Io, Stanton Peale and colleagues had published a paper predicting that Io must be heated by a unique process. Io is so close to Jupiter that the tidal force of that planet's gravity squeezes and stretches Io, slowing its rotation so drastically that, like our Moon, it is locked with one face pointing constantly toward Jupiter.

In the case of the Earth, the Moon raises the ocean tides because its gravitational field, although small, is still enough to drag the Earth's oceans. At the same time, the Earth's stronger gravitational pull tugs on the Moon; long ago, as described earlier, this pull slowed down the Moon's spin until it kept one face constantly toward the Earth.

If Jupiter's picture were this simple, Io would simply orbit around Jupiter as our Moon does around the Earth, with its shape permanently distorted by Jupiter's gravitational field, and there would be no heating of the moon. However, the *other* moons of Jupiter, especially Europa, its nearest large neighbor, tug on Io occasionally and disturb its orbit slightly. This is just enough to pull Io in or out a little away or toward Jupiter, increasing or decreasing Io's tidal force from Jupiter. This causes Io to heat up, much like a strip of metal if it is flexed rapidly.

Scientists are rarely fortunate enough to see their predictions proven so quickly, and even the originators of this theory were a little startled by the dramatic verification of their ideas. Now we know that Io is a moon on which volcanoes erupt almost constantly, sending gases into space (the source of that planetary "doughnut") and heavier material back onto the surface, coating Io in its garish colors. There are entire lakes of black liquid sulfur there, and it is quite likely that, during the history of this bizarre world, the entire moon has been recycled several times, its interior spewing out into space through volcanoes and falling back to coat the surface.

Galileo will test the theory of Io's tidal vulcanism, returning (we hope) spectacular close-ups of the lakes and volcanoes of this extraordinary world.

The same process that heats Io also warms Europa, though more gently. Europa, being farther from Jupiter, doesn't experience as big a gravitational pull, so it doesn't have volcanoes; yet its interior is warm enough to keep water liquid.

With liquid water, the possibility of life inside Europa becomes real. On Earth, all life is directly or indirectly dependent on sunlight, which would be absent underneath the ice cover of Europa. But even on our own planet, we have recently found unique life forms in the perpetual darkness of the bottom of the Pacific Ocean, thriving on an ecology unlike anything else on this planet. Tubeworms and giant clams get their food and energy from natural hot-water vents on the bottom of the sea—undersea geysers. Relying on the chemical energy of hydrogen sulfide rather than solar energy, they are radically different from all other earthly life. If such completely unexpected life forms can exist right here, we should not rule out the possibility that Europa might be a giant fishbowl with all kinds of exotic creatures swimming inside it.

Galileo won't be able to peer inside that fishbowl, but it will pass close enough to give us a much better measurement of the gravitational field, which should show whether, indeed, Europa contains a liquid ocean that some future submarine probe could explore.

Getting *Galileo* Off the Ground

The *Galileo* spacecraft is the most sophisticated robotic space probe ever built. Although it is ready to go, getting it to Jupiter is going to be a problem. It was designed to fly on the Space Shuttle, but in order to move it out of the Earth's gravitational pull, it needs an additional booster of its own.

Originally, NASA decided to use a Centaur rocket booster, which had never been flown on board the Shuttle before. This posed new dangers, because the Centaur is a liquid-fuel rocket that could explode itself, and it would be placed in the cargo bay of the Shuttle, where, if something went wrong, it could blow the Shuttle apart. NASA originally estimated that hazard as slight, until the *Challenger* blew up, showing that their estimates of the Shuttle's safety were much too optimistic. As a result, NASA threw out the design. The Shuttle Centaur was killed, and NASA began looking for other alternatives.

They studied several possibilities, such as using smaller solid-rocket boosters that have flown in the Space Shuttle or other spacecraft before, and are more trustworthy. Unfortunately, such smaller boosters don't have enough energy to accomplish what the Centaur would have: a straightforward launch from Earth to Jupiter, with the orbiter and probe together. One new possibility would be to use two separate launches, one for the probe and one for the orbiter.

Because of continuing budget crunches, NASA has been forced to stop its old custom of doubling up spacecraft. In the good old days, spacecraft were often sent in pairs, such as the two *Pioneers* and the two *Voyagers*. When one instrument on one spacecraft failed, the other one could compensate.

Now, NASA can usually afford only one spacecraft at a time, even though the second one would cost only a small fraction of the first, since the research and development costs remain the same for two as for one. In a sense, NASA and the aerospace industry have become the victims of their own success, since none of the major interplanetary probes of the last decade has failed, even though individual instruments have. Most of the instruments on the *Pioneers, Mariners, Vikings,* and *Voyagers* have continued to function way past their designed lifetime.

So, it seems we don't really need the extra spacecraft. Still, if *Galileo* fails somewhere between Earth and Jupiter, then a decade of work and a billion dollars will have been lost.

The bright side of the budget crunch is that experts are developing new techniques to use the gravity fields of planets and moons like slingshots to swing spacecraft into paths previously impossible to achieve with small amounts of fuel. The *ICE* mission to Comet Giacobini-Zinner did just this, with the most complex gravitational-slingshot maneuver yet attempted. The *Galileo* mission will use even more tricky multiple flybys of the four large moons of Jupiter to move up and down and in and out, like a bumblebee on a bush of four roses. In effect, the moons of Jupiter will themselves be *Galileo's* fuel.

The Delta VEGA technique, mentioned as a possible maneuver for Mars missions, was one intensively studied for *Galileo*. Using this method, the spacecraft would travel out through the asteroid belts and then come back past the Earth, before flying on to Jupiter. One disadvantage: it adds years to the travel time.

A surprising possibility was a Russian launch. To the astonishment of many, the Russians offered to let us use one of their rockets to launch *Galileo*. However, it looks as if it would be as difficult to modify *Galileo* to fly in a Russian spacecraft as it would be to use one of the other approaches, so it's doubtful that much time would have been saved. There was also a reluctance on the part of American egos to have their spacecraft launched by a Russian vessel, especially when our space program was in such a mess.

Another possibility, a recent discovery, was to go in the opposite

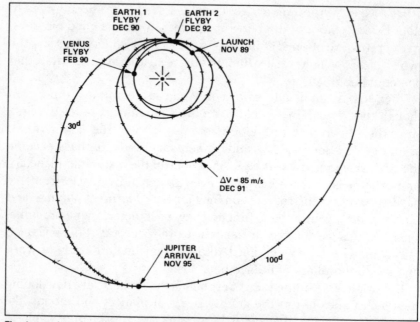

The ingenious technique that will probably take *Galileo* to Jupiter. A new class of trajectory designed by JPL's Roger Diehl, this saves much fuel by flying by Venus once and Earth twice. Ironically, it was discovered by using the same computer programs originally designed to allow *Galileo* to visit Jupiter's four major moons by using their gravitational fields to drastically reduce fuel consumption. It will fly by asteroids Gaspra (in 1991) and Ida (1993).

direction toward the Sun, swinging by Venus and picking up speed that way, then moving out toward Jupiter. This method also adds years, compared with the Centaur approach, but it is the one that will actually be used.

However it gets to Jupiter, *Galileo* will be a fabulous explorer. The orbiter contains special electronics that have been hardened to protect them against the dangers of the radiation belts. It should survive several years in the environment of Jupiter. It will return close-up photos far better than even the superb pictures of the *Voyagers*, uncovering some of the mysteries of the grooves of Ganymede, the bull's-eye of Callisto, the possible ocean within Europa, and the volcanoes on Io.

The delivery of the probe will be a major milestone in the exploration of the solar system. Never before have we actually touched any of the outer planets. The probe will plunge into the atmosphere faster than the Shuttle flies. It will radio back to the orbiter, which will relay the data back to Earth.

The atmosphere of Jupiter is by far the harshest environment ever explored in the solar system. Jupiter's huge gravitational field, with a force of three gees, will accelerate the probe tremendously.

An aeroshell, similar to the coatings that protected reentering astronauts in the *Apollo* programs, will slow down the cone-shaped probe. In a few minutes, a parachute will eject to slow it down even more. Minute by minute the probe will sink into the thick atmosphere of Jupiter, getting hotter and suffering more pressure as it goes deeper and deeper, measuring the temperature, pressure, and light intensity as it sinks.

In the cloud layer that we can see from Earth, the temperature and pressure are not very different from Earth's, but those numbers increase rapidly with depth. Jupiter is extremely hot inside, apparently with left-over heat from the formation of this planet—so hot that it actually radiates about twice as much energy (in the form of infrared light) as it receives from the Sun. With luck, the probe should survive about an hour in the brutal environment.

There will unfortunately be no camera on board this probe, unlike the orbiter. It was difficult enough to put the instrumentation in to measure the atmospheric properties, so something had to be sacrificed. There wasn't room for a television camera, its transmitter, and its data-hungry appetite. (TV requires huge amounts of data to be transmitted, extremely difficult with a low-power transmitter and a small antenna.)

I expect the data from the *Galileo* mission will help us design a future spacecraft that can take a television camera into the atmosphere of Jupiter, floating on a balloon, and thereby surviving for a much longer period. Such a spacecraft could fly within the next 25 years. Imagine what wonders we might see: cloud formations the size of planets, lightning storms greater than any seen on Earth. And with a microphone, what strange sounds might we hear . . . ?

Because conditions on Jupiter at the cloud tops are so similar to the primitive Earth's, some scientists have speculated that life could arise there. Most scientists think the conditions too hostile for life, but at least theoretically there could be creatures there, ones that can avoid sinking into the deep hot layers, which would surely destroy life. It's possible, for example, that there could be life forms resembling hot-air balloons, floating like jellyfish in the upper temperate layers of the atmosphere. There could be winged creatures that fly continuousy through the clouds of Jupiter. Such creatures might even ride the currents of those vast storms, soaring from one place to another high above the hot interior.

It will be a long time before humans venture to the vicinity of Jupiter. But one day, I suspect, astronauts will go there, shielding themselves from the radiation with magnetic fields. They may explore the icy worlds of Ganymede and Callisto, may plunge into the ocean beneath the surface of Europa, and they may see for themselves the sulfur lakes and erupting volcanoes of Io.

Ringworld: Saturn

Beyond Jupiter lies what may be the most beautiful planet in the solar system: Saturn. We have romanticized about its rings ever since Galileo first turned his telescope on this world. Scientists have brooded over their significance for centuries. It wasn't until the 1800's that the great Scottish physicist James Clerk Maxwell proved that the rings could not be solid but had to be made up of countless tiny moonlets. A solid ring, he pointed out, would have been unstable and would have drifted until it crashed into the planet.

A little smaller than Jupiter, Saturn is nine times the diameter of Earth, a hundred times our planet's mass. *Pioneer 11* was the first spacecraft to visit this world, radioing back the first close-up pictures of the rings and outer clouds. It also took a fuzzy picture of Saturn's largest moon, Titan, the only moon then known to have a substantial atmosphere. (Since then, Neptune's moon, Triton, has also been found to have one.) *Pioneer* found that, as with Io, gas escapes from Titan, forming a giant invisible doughnut around Saturn—ironically, making a ring about ten times the diameter of those we see.

Saturn and Titan will be the target of the next major outer-planet probe after *Galileo,* if NASA and the European Space Agency have their way. Right now, the funding is not available, but many scientists on both sides of the Atlantic feel that Titan deserves high priority on our agenda of exploration.

The probe they have proposed is called *Cassini,* after the French-Italian astronomer who discovered the major division in the rings of

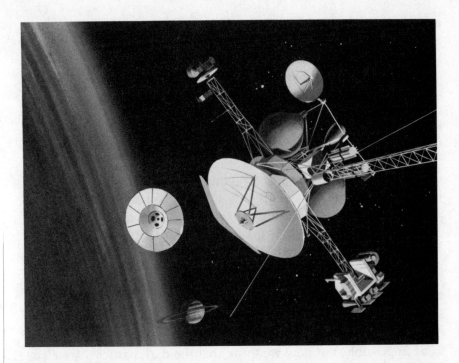

A possible *Mariner Mark II* spacecraft, *Cassini*, that would drop a probe into the atmosphere of the mysterious Saturnian moon Titan, a cloud-covered world that may be coated with complex organic chemicals.

Saturn. *Cassini* would be similar to *Galileo,* with an orbiter that would circle the planet, exploring its atmosphere and its moons, trying to answer some of the questions raised by the *Voyagers*. It would also bring a probe that would enter the atmosphere of Titan.

The *Voyagers* took exquisitely detailed pictures of the rings of Saturn, showing thousands upon thousands of rings within rings, like the grooves of a celestial phonograph record. The study of such planetary rings is important to the understanding of how the solar system formed. Saturn and its rings resemble, like a microcosm, our infant Sun and its satellites.

Data from the *Voyagers* are helping us unravel some of the mysteries of the rings: Where did they come from? Why are they there? Why is Saturn the only planet with a major set of rings? Jupiter and Uranus have less impressive rings around them, and Neptune seems to have at least partial rings, but only Saturn has bright rings easily seen through a small telescope on Earth.

It's starting to look as if rings are a natural phenomenon in the solar system, but we still don't know exactly where they came from. Are they the remains of a satellite that broke up in a collision with another object? Are they, perhaps, the debris left over when the solar system formed, orbits where satellites couldn't quite get together? How do they affect the atmosphere? Recent analysis of the *Voyager* data shows that particles from Saturn's rings "rain" down onto the planet, causing a dark band in the clouds. By tying together all the data from the Jupiter, Saturn, and Uranus flybys, we're beginning to get some insights into these questions, although there are still no clear-cut answers.

One thing Saturn has taught us is that certain satellites act as "shepherding moons"—pairs of tiny moons that gravitationally herd into rings some of the particles that might otherwise dissipate.

To the astonishment of astronomers, the *Voyagers* showed that Saturn's rings have dark markings on them like spokes, which rotate as if they were solid, even though the rings are made of many particles moving in different orbits at different speeds. Spokes seemed to be impossible, but now we think that the dark particles are electrically charged and are caught up in the magnetic field of Saturn. Charged particles tend to be swept around as if they were frozen to that magnetic field, and this could explain why those spokes can last for days.

Perhaps this obsession with understanding the motion of particles seems unimportant, but that phenomenon is precisely what we must study to understand how the solar system formed. Indeed, we think that every large object in the universe was originally formed by gas and dust coming together, attracted by gravity, forming small clumps that then attracted other clumps and became bigger and bigger, giving rise to the planets and stars and whole galaxies. Planetary rings allow us to see this process in operation.

Saturn itself turned out to be a rather featureless planet, compared with Jupiter. There were cloud bands, and indications of storms, but nothing as dramatic as on Jupiter. But Saturn's moons turned out to be fascinating worlds in themselves. Saturn has only one large moon, Titan, comparable in size to the four large moons of Jupiter, but it has a half-dozen medium-sized moons, a few hundred miles in diameter, with strange features discovered by the *Voyager* cameras. One of them,

*What is the origin of the devouring curiosity that drives
men to commit their lives, their health, their reputation,
their fortunes, to conquer a bit of knowledge, to stretch our
physical, emotional, or intellectual territory? The more I
spend time observing nature, the more I believe that man's
motivation for exploration is but the sophistication of a
universal instinctive drive deeply ingrained in all living
creatures. Life is growth—individuals and species grow in
size, in number, and in territory. The peripheral manifestation of growing is exploring the outside world. Plants
develop in the most favorable direction, which implies that
they have explored the other orientations and found that
they are inadequate.*

*Some plants send feelers at great distances; they send
avant-garde shoots before they invade the space that has
been acknowledged propitious. For young animals the world
is to be explored and discovered from their birth on, and
that exploration only ends with death; for the young fox,
wilderness is unlimited; for a tuna, the oceans are infinite.
Still in the animal world, the physical need for exploration
develops as well in individuals as in collectivities—tribes,
schools, swarms, packs. In fact, if the baby human being
shows the same motivation as a young cat, to explore with
all his sensors the strange environment he was born into,
the big difference is that the little baby soon stands erect.
That radical change came in evolution the day described so
well by Ovid, a few years after Christ was born. "God
elevated the forehead of Man," wrote Ovid, "and ordered
him to contemplate the Stars."*

—Jacques Cousteau

Mimas, looked like the Death Star out of the *Star Wars* film, with its
huge crater almost as big as the moon itself.

The most exciting find in the Saturn system was the close look at
Titan. I remember very well the approach of *Voyager 1* to Saturn. We'd
known since World War II that Titan had an atmosphere, but early
measurements had shown it was quite thin.

It was a big puzzle in astronomy that a moon should have any atmosphere at all. After all, our own Moon, somewhat smaller than Titan, has no atmosphere because its gravitational field is too weak to hold onto any gas for any long period of time. Why then did Titan have an atmosphere? Shouldn't it have leaked off into space during the billions of years of solar system history? We eagerly waited at JPL to see what the spacecraft would see when its camera swung to Titan.

When the time arrived, what we saw was nothing! The planet was covered with a thick, orangish cloud that prevented us from seeing the surface. Most scientists were, at first, terribly disappointed.

Titan had an astonishing 1.6 atmospheres of pressure—60 percent greater than the air pressure you are now breathing, if you're on Earth. This thick atmosphere means that Titan is actually a much more interesting world than we originally thought.

Titan's atmosphere, like Earth's, turned out to be mostly nitrogen, which is difficult to detect from Earth. (That explains why it hadn't been seen before.) Titan's clouds are something like the smog above Los Angeles. Theoretical models of that moon's chemistry show there's likely to be ethane and acetylene, organic (that is, carbon-compound) gases that can become liquids at the temperatures and pressures on Titan. So there may be rains and snows and lakes, even oceans of these organic chemicals there. The oceans may be of ethane and methane, and the reddish-brown haze may consist entirely of complex organic chemicals.

The temperature on Titan is very cold, around 300 degrees Fahrenheit below zero. This probably prevents truly complicated organic chemicals from forming, but over the billions of years of solar system history, it is quite possible that strange chemicals have arisen there. It is even remotely possible that some kind of life could have come into being, particularly during the early, warmer days of the solar system. Even if there is no life now, the chemistry and meteorology of its atmosphere and surface must be complex and unique, well worth exploring. So Titan is a strange, exciting world, much more accessible than the giant planets, ripe for the *Cassini* mission.

Though not fully approved, *Cassini* has a good chance of being launched in 1996 as a joint NASA/European Space Agency mission. Its radar would penetrate Titan's clouds, and a probe would parachute onto Titan in 2002. The probe would analyze the moon's atmosphere and ocean. The next generation of spacecraft after that might fly through the atmosphere on a balloon, as the Russians and French will do on Venus. In some ways, this would be easier to accomplish on Titan than on Venus: A cold temperature is much easier on electronics.

It would be fabulous to have a camera floating above Titan, surveying its exotic landscape as the chemical oceans slosh against the unfamiliar floor of this alien world, flying over a landscape coated with billions of years of organic deposits. Such a mission will probably happen in our lifetimes.

The Risky Voyage toward Neptune

The next planet out is the strange world of Uranus, tipped on its side like a discarded toy top. *Voyager 2* is the only spacecraft that has visited it, the only one likely to in the next 25 years.

Although its dark rings help us understand Saturn's, and its small moons, with their crazy-quilt of surface features, mystify geologists, it is so far away and difficult to get to that no one is proposing to return there soon.

As more sophisticated methods of propulsion are developed and electronics becomes ever more compact and cheap, we will probably see the time when tiny, inexpensive probes can be shot all over the solar system; that may be when Uranus will be revisited. Until then, we must content ourselves with the one remaining scheduled first visit in our current exploration of the solar system: Neptune.

Since *Voyager 2* survived so well its encounter with Uranus, we are optimistic that it will sail on to Neptune without a major hitch. We'll get our first data back from Neptune in 1989.

Neptune is half again as far away as Uranus, presenting difficulties of darkness and weak signals similar to the Uranus flyby, but NASA upgraded its antennas and receivers, and developed new techniques to process the data on board the spacecraft.

There are two major targets for this Neptune mission: the planet itself and its moon Triton. It has one other moon, called Nereid, but that's a tiny one of no great importance, as far as we can tell, compared with the extremely interesting moon of Triton.

The main problem we face is Neptune's partial rings. Apparently they are something like the rings around the other planets, but they have not formed complete circles the way those of Jupiter, Saturn, and Uranus have. They present a real hazard to the spacecraft, since NASA wants to take *Voyager 2* close over one of the poles of Neptune and very close to Triton. In order to do this, *Voyager 2* must fly close to one of the fragmentary rings. Whether or not it can will depend on more precise measurements of these strange rings. It was hoped that the Hubble Space Telescope would be in Earth orbit before *Voyager* got to Neptune, in which case we'd have had good shots of the partial rings to help plan this mission, but it couldn't be launched in time.

Triton resembles Titan in that it has a substantial atmosphere of nitrogen. It, too, may have lakes or even oceans of liquid nitrogen. What chemical reactions may have occurred there during the billions of years of solar system history are yet to be discovered. Liquids, as I've described, were essential to the evolution of life on Earth, but our temperatures, here in the comfortable inner solar system, were a lot warmer than those on Triton. So until recently, most scientists agreed that in the frigid realm of that moon, the complex reactions hospitable to life would be unlikely.

However, we have now come to realize that chemical reactions still take place even at very cold temperatures. When the temperatures are truly frigid (as in liquid hydrogen, for example), only a few degrees above absolute zero, the familiar domain of chemistry is transformed into the surreal world of quantum physics. Quantum theory, the basis of our understanding of the world of atoms, is a world where a particle can pass through an obstacle—"tunneling" through it as if it were not there. This is a world where particles are not hard, rigid, well-defined little balls, but fuzzy objects that defy intuition. It is a world where absolutes change into statistics, where you can no longer state that an object is either in a box or not in a box; you can only state that the chances are x percent that it's in the box and y percent that it's not.

Under these conditions, chemical reactions that would have been expected to slow down to nothing can actually continue. Chemicals can form by this tunneling process much faster than by conventional means. During billions of years under such conditions, who knows what strange phenomena have taken place on Triton?

Pluto and the Shadowy Beyond

After Neptune (ordinarily) lies Pluto. By chance, Pluto at this moment happens to be slightly inside the orbit of Neptune, but for most of its 248-year period, Pluto spins outside the orbit of Neptune. Pluto is probably a former satellite of Neptune, which would explain its strange orbit. There are no spacecraft planned to go to Pluto, a small world a fifth the diameter of Earth, about which little is known. It is a world where the moon Charon is a quarter the size of its parent planet—about the same ratio as between Earth and our Moon.

Our one hope for getting more information about this planet and its moon comes from the Hubble Space Telescope, which from Earth's orbit will give us a far better picture than we could ever achieve through the murky atmosphere of Earth itself.

Beyond Pluto there may be other planets—undiscovered worlds. There are puzzling discrepancies in the motion of Neptune and Pluto over the past century that, if real, must mean the existence of at least one other planet. Sometimes calling it Planet X, scientists have hunted it for decades without success.

Because it would be so far from the Sun, a planet beyond Neptune and Pluto would be very dim. It would be hard to detect against the background of billions of faint stars. And if it's not in the same plane as the orbits of the other planets, it could be many years before we find it.

But there's no reason to think that the solar system suddenly ends with Pluto. To think this would be to make the same mistake ancient astronomers did before the planet Uranus was detected. We simply don't know much about what orbits around the Sun outside Pluto's path. There could be all manner of exotic bodies there.

At the very least, we believe there are billions of billions of comets slowly orbiting in a complex dance around the Sun, waiting for that passing star that might fling them into a fiery death among the planets of the inner solar system.

At this moment, probing the mysteries beyond the edge of our solar system, the tiny *Pioneer 10* spacecraft is taking our first slow steps into the interstellar unknown. It should continue to transmit data back for the next ten to twenty years. Its radioactive power supply is declining, and its tiny transmitter—with about the power of a refrigerator light bulb—becomes harder and harder for the most powerful receivers and antennas on Earth to pick up.

Pioneer 10 and its brothers, *Pioneer 11* and the two *Voyagers*, were not designed specifically for interstellar space, so the data they send back will be rather limited. Primarily, we will have measurements of the dust, the magnetic fields, and the gases of interstellar space. If there is an unknown planet out there, and if one of those spacecraft passes anywhere near it, we will be able to detect it by the change in the spacecraft's path as recorded in subtle changes in its radio signals. Planet X, if it exists, may be found by our interstellar ambassadors.

Both *Pioneers* and both *Voyagers* are still working well as they venture into the edge of interstellar space, boldly going where no spacecraft has gone before. If all goes well, they'll continue to send data back as far ahead as the 21st century, by which time we'll probably have launched the first probes deliberately designed to explore nearby interstellar space.

TWELVE

Glimpsing Infinity

THE UNIVERSE

Beyond Neptune and Pluto, in the blackness of the universe, unknown things lurk, sometimes betraying their presence like animals in the jungle night, by a faint radio squeak or a soft glow of light.

There are many things we know now about interstellar space and beyond, primarily from Earth-based observations and from Earth-orbiting satellites, and the next decade will see more powerful observatories on Earth and in space than any yet made. Two telescopes bigger than any now in existence will be built in the U.S. In addition,

> "It is a most beautiful and delightful sight," exclaims Galileo, in describing the discoveries he had made with his telescope, "to behold the body of the moon, which is distant from us nearly sixty semi-diameters of the earth, as near as if it was at a distance of only two of the same measures. . . . And, consequently, any one may know with the certainty that is due to the use of our senses that the moon assuredly does not possess a smooth and polished surface, but one rough and uneven, and, just like the face of the earth itself, is everywhere full of vast protuberances, deep chasms, and sinuosities."
>
> —*Garrett P. Serviss*, Astronomy with an Opera-Glass, *1888*

several astronomical satellites should give us some fabulous views of the universe beyond our solar system.

The European Space Agency is building *Hipparcos*, the first satellite designed specifically to measure the positions of the stars. It will give us an unprecedented map of the stellar neighborhood in which our Sun wanders, will probably detect nearby stars not noticed before amid the clutter of the heavens, and it will most likely find planets around other stars.

NASA has several superb orbital instruments planned. The Advanced X-ray Astrophysics Facility will spy on explosions in the cosmos, black holes, and the strange powerhouses on the edge of the universe known as quasars. The Cosmic Background Explorer will make delicate measurements of the microwave sizzle left over from the Big Bang that formed the entire universe.

And the blockbuster of them all will be the Hubble Space Telescope.

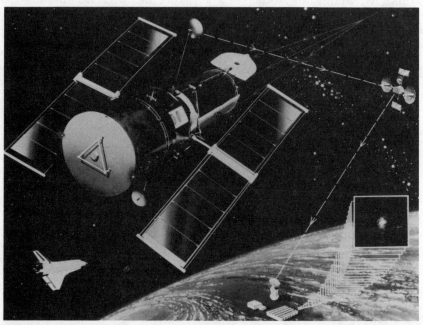

The Hubble Space Telescope, already built, a 94-inch mirror to be launched by the Shuttle. It should be able to see farther into depths of space than ever before, to the edge of the cosmos, where light has been traveling for billions of years—since shortly after the birth of the universe.

For the first time in the 30-year history of the space age, we will have a large (94-inch diameter) astronomical telescope in space. Out beyond the murky atmosphere that has plagued every telescope observer since Galileo, we will see closer to the edge of the universe and farther back in time than ever before. We will see very nearly to the time when the universe began.

The Geography of the Heavens

We live in a universe so vast that most people cannot begin to comprehend its size.

Pluto never gets farther than 50 times the Earth's distance from the Sun—five billion miles—but the distance to even the nearest star beyond the Sun is so vast that it must be measured in light-years, the distance that light travels in one year. Since light moves at 186,000 miles per second, one light-year is six trillion miles. The nearest star to the Sun is a triple star system known as Alpha Centauri, at a distance of about four light-years away. By comparison, Pluto is less than one-tenth of a percent of a light-year from the Sun.

So even though it takes us years to get to the planets of our solar system, such distances are insignificant compared with the scale of the universe. At the speed of our four interstellar spacecraft, it would take tens of thousands of years to reach Alpha Centauri if they were aimed there (which they're not).

The Sun is just one of a swarm of several hundred billion stars we call the Milky Way galaxy. On a clear night, away from city lights, the faint band of light we see in the sky is our view of this galaxy in which we live. The Milky Way is shaped like a discus, but because we live inside this disc, when we look out we see thousands of distinct stars that seem to be randomly scattered all over the heavens. These are, for the most part, the nearby stars.

Our Milky Way galaxy is 100,000 light-years in diameter. Our solar system is about 30,000 light-years from the center of the disk—two-thirds of the way out toward the rim. We are in the suburbs of the galaxy.

So far away is the Galactic Center that if there is a civilization there looking at our Sun and Earth with unimaginably giant telescopes, all they will see on Earth are the natural features of the planet. They will be seeing us as we were 30,000 years ago, because light from the Earth

Many cosmic phenomena have only come to be recognized in the past thirty-five years, largely through the introduction into astronomy of radio, X-ray, infrared, and gamma-ray techniques. None of the new phenomena had been anticipated before World War II, and it is natural to wonder how many more remain unrecognized even today, how rich and complex the universe might be. Further, if technological advances already have helped us uncover so many new cosmic features, how many more innovations of similar kinds could we put to use in future cosmic searches?

—Martin Harwit, Cornell University

has taken that long to get there. This was well before the pyramids of Egypt were built. Perhaps they are now watching mammoths and saber-toothed tigers roam the Earth, as the newly triumphant species Cro-Magnon spreads across the world, having just conquered its predecessor, the Neanderthals. The first Americans have just crossed from Asia onto the continent of North America and are spreading southward.

Despite the vastness of our galaxy, it is tiny compared with the distances of the universe at large. There are at least ten billion other galaxies in the universe like our Milky Way. Some are bigger, some smaller. The nearest is the Large Magellanic Cloud, a small galaxy discovered by the explorer Ferdinand Magellan in the 16th century.

To Magellan it was a small patch of light in the night sky, visible only from the Southern Hemisphere. This is where the first "nearby" supernova in 383 years recently was seen. It's about 170,000 light-years from us, and that's our *nearest* galactic neighbor! The nearest galaxy similar in size to our own Milky Way is the Andromeda galaxy, a fuzzy patch of light also visible to the naked eye, two million light-years away. And even this distance is small, compared with the size of the universe.

The most distant objects ever detected—quasars that may be exploding galaxies—are more than ten billion light-years away. Despite the immensity of our universe, we have learned a remarkable amount about it. Thanks to telescopes that can look not only at light, but also at infrared, ultraviolet, radiowaves, and other types of radiation, we are able to learn a great deal about this place in which we live.

We are fortunate that as every body in the universe sends out tiny pieces of itself, in the form of lightwaves, radiowaves, and fast-moving particles such as electrons and protons, astronomers can capture some of these pieces of distant stars and galaxies and decipher them, using logic worthy of Sherlock Holmes to deduce the nature of the beasts from which they came. Conventional telescopes—starting with

(a)
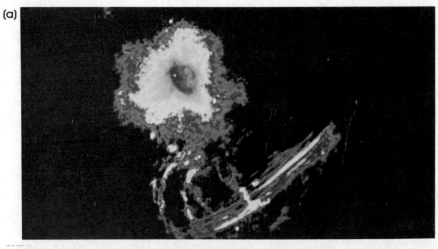

(b)
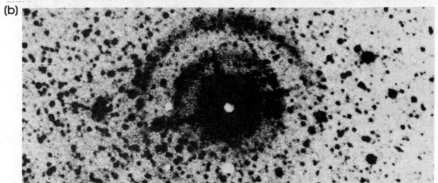

(a) VLA radio image of the center of our Galaxy, showing strange arcs up to 100 light-years in size, probably caused by magnetic fields.

(b) Supernova 1987a, the first major one in centuries, showing a remarkable "light echo"—two rings around it, from dust in distant space. (Negative.)

Galileo's small, hand-held instrument, ranging up to Mt. Palomar's 200-inch-diameter mirror and the larger but less-effective Soviet Zelenchukskaya 240-inch mirror—have shown us the basic characteristics of the universe. They have shown us pictures of many of the important objects out there, from stars to galaxies.

But much of the universe is invisible. There are pulsars and black holes, and special types of galaxies that do not radiate much visible light. These we can only detect with other types of instruments.

The Death Star: The Nemesis Theory

Somewhere between Pluto and Alpha Centauri there may lie another star, sometimes called the Death Star. Its existence has been deduced indirectly by studying the history of the Earth. Intimations of this star's presence first arose a few years ago when Luis Alvarez, a Nobel Prize-winning physicist, along with his geologist son Walter and colleagues, was wondering about a strange layer of clay found on a hillside in Italy. Less than an inch thick, the layer contained an unusual amount of the rare metal iridium.

Alvarez knew of only one place where iridium is relatively common—in outer space. Meteorites often contain unusual amounts of this metal. He suggested that perhaps the clay had been laid down following the collision of a giant meteorite with the Earth. (Later, it was suggested that a comet could have had the same effect.) The fascinating thing about this layer was that it dated back to the time when the dinosaurs died out—65 million years ago. It was then that something wiped out many species on Earth, including the dinosaurs.

Scientists have been fighting over this theory ever since. The evidence, in my opinion, favors Alvarez, but there's still a substantial number of geologists who think widespread volcanic activity or some climatic change did in all those species.

Two other scientists, David Raup and John Sepkoski, began to think about other mass extinctions that have occurred in the history of the Earth. The geological record shows a couple of dozen periods when many species were wiped out simultaneously. They studied the timing of these extinctions and published a paper claiming that many of them occurred at regular intervals, roughly every 30 million years.

Their claim was even more controversial than the Alvarez theory.

If correct, it meant that we had to look not only for a single catastrophe to explain the death of the dinosaurs but for some recurrent event that decimated life on Earth every 30 million years or so.

A number of ideas came up, but none seemed to hold water until Richard Muller, Marc Davis, and Piet Hut suggested that the Sun might have a dim, heretofore undetected companion orbiting around it, which Muller named Nemesis. If the Sun has a companion star—as most stars do—and if this companion is in an orbit about halfway between us and Alpha Centauri, then it will take about 30 million years to complete one orbit around the Sun. If the orbit is very elongated, that star will spend most of its time at a remote distance, on the order of one or two light-years away, and for only a relatively short period of time will it be near the planets.

It doesn't actually have to get close to the planets to have a terrifying effect on the solar system. All it has to do is pass through the Oort Cloud of comets that orbits beyond Pluto around the Sun. Its gravitational field would stir up those bodies like a stick in a hornets' nest, and a storm of a million comets might come raging into the solar system; some of them might well smash into the Earth.

One of the beauties of this theory is that it can be *tested*. If Nemesis exists, we should be able to find it within the next couple of years. There are Earth-based telescopes looking for it now, and if they don't find it, the European Space Agency's *Hipparcos* should. If nobody finds it, we'll have to throw the theory out.

A nice by-product of this search is that it should uncover many other nearby stars currently unknown. We might even have neighbors other than Alpha Centauri that would be easier for spacecraft to reach.

The more important fundamental laws and facts of physical science have all been discovered, and these are now so firmly established that the possibility of their ever being supplanted in consequence of new discoveries is exceedingly remote.

—Albert A. Michelson, Nobel laureate, in 1902

Interstellar Nurseries and Supermarkets

Moving beyond the nearest stars, we find all kinds of other objects in our galaxies. There are almost-spherical clouds of gas called planetary nebulae, produced when a star is dying. There are other irregular clouds in which stars are being born. From birth to death, we can see the history of our own solar system written in the sky. The stellar nurseries are hard to see clearly with visible light because they are often clogged with dust, the dust out of which the stars and planets were formed.

However, the dust glows strongly in infrared light, which is redder than the reddest light we can see with the human eye. Infrared is usually produced by warm objects, which is how military sniperscopes see in the dark. Astronomers watch the infrared glow of dust clouds in space.

The problem—so often the case with astronomy—is that our atmosphere filters out much of the radiation from the sky. In fact, the reason visible light is visible is simply that over the billions of years of evolution, eyes evolved to use whatever radiation was plentiful. Sunlight was the only game in town. If the Sun had been cooler, we would see by infrared; if hotter, by ultraviolet.

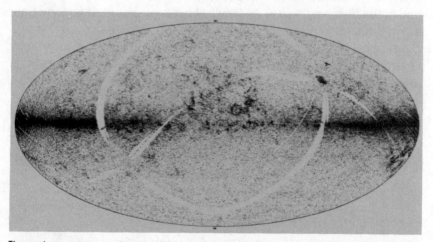

The universe as seen by the *IRAS* infrared telescope. The Milky Way fills the middle horizontally. Many of the sources are giant molecular clouds, or galaxies not visible to ordinary telescopes. (This is a negative, in which the bare swaths are where data were not available.)

In order to see the infrared universe, England, Holland, and the U.S. built a satellite called *IRAS*—the Infrared Astronomical Satellite, launched in 1983.

It functioned for almost a year, and among its findings was the startling discovery that rings of gas and dust orbit around many stars. The bright nearby star Vega, 25 light-years from us, turned out, to the surprise and delight of astronomers, to have a ring circling around it, visible only in infrared. The ring seems to be made up of dust or possibly comets, debris similar to that out of which the solar system formed. This finding strongly suggests that the formation of planets is a normal event in the history of the universe, not a freak event unique to our Sun. Thus, many of the stars we see at night probably have planets orbiting them, and some of those planets may well harbor life.

NASA now proposes to build a significantly enhanced version of *IRAS*, called *SIRTF* (Space Infrared Telescope Facility). This will be an even more powerful interplanetary telescope carried by the Space Shuttle to find out more about the infrared universe.

The Europeans are also designing a similar spacecraft, and some scientists are wondering whether the Europeans and Americans can get together to design one facility that will do most of what both groups would like. In these days of limited budgets, such cooperation will probably be a necessity.

These infrared observatories will also explore a discovery that has profound implications for the existence of life around other stars: giant molecular clouds. These are vast clouds of gas and dust, with huge amounts of large molecules. Many of them are complex organic compounds similar to the ones out of which we are made.

A typical giant molecular cloud contains hundreds of thousands of times as much matter as our Sun, although most of it is hydrogen. The existence of these clouds proves that the universe makes the basic molecules of life in great abundance, giving added incentive to those who search for life elsewhere, and raising the question: Could life have been seeded when the Sun passed through such a cloud?

Whether or not that was the case, it's clear that the galaxy is a factory for the manufacture of complex molecules. The best way to study these clouds is by using a combination of infrared and radio astronomy, both on Earth and in space.

As usual, the Earth's atmosphere interferes with the observation of many of the infrared and radio (microwave) frequencies we need to see. The ideal microwave radio observatory would be in space, and orbiting dishes have been devised that would allow us to observe these microwaves in their full glory. Engineers and scientists have put together designs that can be assembled or unfolded in space. Both the Soviet Union and the U.S. are intent on placing radiotelescopes in space, and the Russians expect to have one completed in the 1990's.

From orbit we'll see far more clearly these strange supermarkets and nurseries of space, and detect even more complex molecules, giving further insight into the origin of life and the possibility that we are not alone in the universe.

Black Holes and Other Weirdness

The radio astronomer leads a very different life from his or her optical counterpart. While most optical astronomers climb up desolate mountains to spend frigid nights in an unheated observatory, the radio astronomer can often do the job in sunshine down in the lowlands. Where the optical astronomer peers through huge versions of the lenses that Galileo used (with mirrors usually substituting now for the lenses of Galileo's telescope), the radio astronomer uses something very much like a satellite dish.

Radio astronomers "listen" to pick up the signals of distant disturbances in the universe. For example, in 1967, a young Irish woman named Jocelyn Bell (now Burnell) discovered strange tickings coming from the sky. She was working in England on a radiotelescope designed by British astronomer Antony Hewish. She observed that the ticks came about once a second, and seemed so artificial that before long they were dubbed Little Green Men. However, these signals turned out not to be aliens transmitting to us, but rather the first proof of the existence of a bizarre object that theoreticians had predicted but which astronomers had never seen—the neutron star.

When a star much larger than the Sun dies, it goes out in a blaze of glory, in a supernova explosion. For a time it may be brighter than an entire galaxy, and when it's finished, what often remains of the original star is a super-compact, rapidly spinning object, the neutron star. As much matter as is in the entire Sun may be compressed down to a tiny ball about ten miles in diameter. It was such a spinning ball of enormously dense matter that Bell and Hewish found.

> [In the unpiloted-spacecraft program] we have seemed to
> allow ourselves to be so concerned about failure . . . that
> we are less innovative, less willing to move out into the
> unknown.
>
> —Philip Culbertson, NASA general manager.

Now called pulsars, neutron stars have graduated from theory to textbook, and several hundred of them have been found.

The pulsar is a place on the very edge of physics as we know it. It is a star where the laws of nature are stretched to the limit, where space itself is warped. Matter is so compressed that one teaspoonful weighs many tons. The pulsar has a magnetic field a trillion times stronger than Earth's, and it slings electrons and protons to energies greater than that of any of our particle accelerators. It is the electrons, screaming in magnetic fields, that the radio astronomer hears, while the pulsar spins around like a lighthouse every second or so, shooting out the beam of radio waves generated by the electrons.

One remarkable object just discovered is sure to be the target for some of the astronomical spacecraft to be launched soon. Astronomers in Europe and the U.S. have found two stars orbiting around each other closer than any two stars ever before observed.

They're closer than the Earth and the Moon, only 80,000 miles apart, and they zip around each other every eleven minutes at almost half a percent of the speed of light. One of them is a white dwarf—a dense, worn-out star with the mass of an ordinary star, shrunk down into an object the size of Earth; the other is a pulsar. Matter is drained off the white dwarf onto the neutron star much as a vampire sucks the blood out of his victim. The pair should tell us much about conditions that test the laws of physics, including the extremely difficult-to-detect radiation of gravity waves—ripples in the fabric of space and time.

One of the most mystifying discoveries of the last two decades may be caused by neutron stars: gamma-ray bursters, first discovered by a secret U.S. Air Force satellite in 1967, but not declassified until 1973.

About once a day, a burst of gamma rays (similar to X-rays, but with higher energy) hits the Earth for a few seconds. The eerie thing about them is that each burst comes from a different part of the sky. None has been identified with a known object. And they almost never repeat! It's as if we are in the dark, surrounded by a crowd, and one person shrieks, followed by silence; then another, and so forth, but never the same person twice. There are many theories about the origin of these bursts, usually relying on neutron stars, but none is generally accepted.

Astronomers need more data to weed out the wrong theories, and in 1990, the U.S./German *Gamma-Ray Observatory (GRO)* is scheduled to launch; its mission will be to examine the radiation from bursters and any other gamma-ray sources in space. In the meantime, Earth-based astronomers are building special detectors they hope will spot flashes of light in the sky from some of the bursters.

GRO will also examine what seems to be, for a change, the *solution* to one of the great mysteries of astronomy. Ever since Victor Hess, an Austrian, discovered cosmic rays in 1912, we've known that the sky is full of energetic particles (electrons, protons, and others) that rain down on us constantly. Similar to radioactivity, but more energetic, the source of these particles' energies has been debated for the better part of a century. Cosmic rays as energetic as a hundred billion billion electron-volts have been detected, and there is no end in sight; the higher the energy, the rarer they are, but so far there seems to be no limit to their power.

Until recently, no one could pin down a particular object as the source of cosmic rays. Finally, a certain gamma-ray emission was traced to an X-ray source, Cygnus X-3. This object, the third- "brightest" X-ray source in the constellation of Cygnus, is hidden behind dust on the edge of the galaxy, but its gamma- and X-rays give its location away. It seems to be a binary in which one of the components is an ordinary star, and the other a pulsar that somehow accelerates matter.

A neutron star is just one step away from the most bizarre of all objects that we know of in the universe—the black hole. A star more than three times as massive as the Sun may end its existence struggling to be a neutron star. But it is so massive that the force of gravity crushes it irresistibly. Even neutron matter cannot withstand such a gravitational force, and it collapses into itself, we think.

No one knows for sure what happens, because the laws of physics as we know them may break down under these conditions. The laws we know seem to dictate that in this case matter becomes a perfect singularity—a place where the density of matter is infinite.

After pulsars were discovered, it became inevitable that black holes, if they existed, would be found. Astronomers have now found a number of objects radiating enormous amounts of energy that seem to require black holes to explain them. They "shine" by radiowaves, X-rays, and gamma rays, and the U.S./Dutch Advanced X-Ray Astrophysics Facility (AXAF) will examine the radiation from these violent objects and search for more in 1995.

The first black hole to be found was probably Cygnus X-1, a binary star in which one star is a supergiant twenty times the mass of our Sun, and the other is invisible, with half the mass of its companion. The invisible one is the black hole, announcing its presence by its attraction of the visible star and by its X-ray emissions.

A black hole is so massive that nothing can escape from it. On a normal object, there is a minimum speed needed to escape the gravitational field, called the escape velocity. On Earth, the escape velocity is 25,000 mph—anything slower just falls back down. But the escape velocity from the surface of a black hole is greater than the speed of light, and since Einstein tells us that nothing can travel faster than that speed, nothing can escape from the black hole. Even light itself, if it starts to leave a black hole, curves back down and hits the hole once again. That's why it's called *black*. Not even light can escape.

This doesn't mean it's undetectable. The gravitational and magnetic fields of a black hole stretch out into space and can pull in gas and stars, and accelerate electrons that generate radiowaves.

Furthermore, the matter flowing down the cosmic drain of the black hole becomes so hot that instead of just emitting infrared radiation, or visible light, or even ultraviolet, it emits radiation stronger than ultraviolet—X-rays. X-rays from space were discovered by rockets and satellites, since the Earth's atmosphere absorbs them. Probably, many X-ray sources in the sky are black holes. AXAF will be a window into places where almost inconceivable forces are ripping apart the very fabric of space and time.

One of the strangest places in the universe turns out to be the center of our own Milky Way galaxy. There appears to be a gigantic black hole right in the center of our galaxy, as massive as a million suns. Our Sun orbits around it, taking a quarter of a billion years to complete one pass.

Radiotelescopes have peered through the dust blocking our view of the Galactic Center, and have found huge amounts of noise coming from it. In the last couple of years, one of the most powerful radiotelescopes has been aimed at the Galactic Center. Near Socorro, New Mexico, there's a field of twenty-seven 85-foot-diameter dishes, known as the VLA (Very Large Array). (Viewers of the movie *2010* saw a glimpse of them at the beginning of the film.)

The first time I saw VLA images of the Galactic Center, I was astonished by the strange structures in the center of our galaxy—enormous lines and curves unlike anything ever seen in space, like highways between interstellar civilizations. However, they are probably nothing more than great currents of super-hot gas, channeled by immense magnetic fields—perfectly natural, though not at all understood. The VLA has given us, instead of the usual scientific plots of radio intensity, details as fine as in a photograph.

Astronomers have recently finished the first successful test of QUASAT, a project combining a communications satellite with Earth-based radiotelescopes, to reveal finer detail of radio sources than could be obtained on the ground. Their success means we'll be seeing more use of spacecraft as radiotelescopes linked to ground-based ones, which will show us what's going on in black holes including the one at the center of our galaxy.

Time Machine: The Hubble Space Telescope

The universe started with a Big Bang, when all matter was compressed to an amount smaller than a pinhead. It exploded in a burst of energy never seen since, and out of the debris formed galaxies, stars, and planets.

That was ten to twenty billion years ago, so when we look with our best telescopes and see the most distant objects billions of light-years away we are seeing the universe as it was in the distant past.

Now that the Shuttle is running again, NASA plans to launch the Hubble Space Telescope, with a 94-inch diameter mirror, by far the largest ever put into space—half the diameter of the Mt. Palomar telescope. It's built and ready to go, probably at the end of 1989.

Hubble will be our greatest time machine. We will see farther into the universe than the best Earth-based telescope can. We should see galaxies and galaxy-like quasars being born, and perhaps phenomena our theorists haven't dreamed of.

It will be the first large optical telescope put into orbit to view the universe. (There have probably been other such telescopes in orbit before, put up there by the CIA and the KGB, but aimed the wrong way, from the astronomer's point of view.)

Deciphering the nature of the early universe is one of the most exciting endeavors of physics and astronomy today. Scientists have found that a simple Big Bang isn't enough to explain the smoothness of the universe, so they've invented an inflationary universe—one that had a sudden, extra bang as it expanded.

Cosmic strings have been envisioned—primitive, vast loops of the pure substance of space that may have been the first objects to form, and that may still lurk among the galaxies.

Particle physicists seem to be close to unifying all the forces of nature into one Theory of Everything. If they're right, the conditions during the first split-seconds of creation determined which particles came into existence and which did not.

Hubble should shed light on all these questions, and on another major mystery of the universe: the case of the missing matter.

Astronomers can calculate how much matter there is in the universe by two methods, which might be called the speedometer method and the accounting method. In the speedometer approach, they measure how fast stars and gas move around the center of the galaxy. The speed of one object orbiting around another—such as the Earth around the Sun—is a measure of the attracting body's mass. That it takes the Earth

© 1986 Newspaper Enterprise Association, Inc.

one year to circle the Sun tells us how massive the Sun is. Similarly, the speed of stars orbiting around a galaxy tells us how massive the galaxy is.

The accounting method, on the other hand, lists all the known objects in a galaxy (stars, gas, dust) and uses our knowledge of how massive each of these items in the inventory is. Then you add up all the masses and figure out how much matter there is in the galaxy.

But the two answers don't agree. A cosmic audit shows that the speedometer method gives a bigger mass than the accounting method. Something's wrong. Something's missing. There must be something in these galaxies we've left out of the inventory. As much as 90 percent of the matter in our own galaxy is mysteriously unaccounted for.

Maybe it's in black holes we haven't detected, or in the form of an enormous number of neutrinos. Perhaps it's a form of matter that has merely been theorized but never actually observed—particles invented by theoreticians and given names like Italian desserts—gravitinos, photinos, and a whole host of others predicted but never seen.

Something must be there. Not only that, but the something is a major part of the universe. It also determines the answer to the ultimate fate of the universe: Will the universe keep expanding forever, or will the Big Bang collapse in a Big Crunch?

Or, as Einstein put it, is the universe open or closed? He was the one who discovered that the universe might be an enormous prison cell from which no escape was possible. He calculated that if the universe had enough matter, then the very fabric of space and time would close in on itself in a higher dimension, in a way that is difficult for us three-dimensional beings to comprehend. The universe would be closed through the fourth dimension.

It is as if an ant were living on the surface of a basketball. He trudges in one direction on the basketball, and it seems to go on forever. In fact, he may come back again and again to the same spot, even though he's going in what seems to him to be a straight line. The poor guy can't see that his whole universe, which looks flat and two-dimensional, is really curved through the third dimension. A mathematically inclined ant would be forced to the conclusion that he was in a closed universe.

We may be just like that ant, but our universe could be closed in a higher dimension. If this is right, then if we took off in a spaceship and traveled in a straight line forever and ever, we would come back

Serendipity means that you dig for worms and strike gold. Happens all the time in science. It is the reason why "useless" pure research is always so much more practical than "practical" work.

—*Robert A. Heinlein, author*

to the place we started from. If we kept on traveling, we would keep coming back again and again. The circumference of our universe would be tens of billions of light-years. In other words, our sphere is of immense size, but it is still finite. It is limited. It's like living on the inside of a black hole.

But, in case you feel claustrophic, the universe may be open. If we don't quite have the critical matter needed to close the universe, then it's open. In that case, we are not living in a prison cell but, rather, we are in a free and boundless universe.

In an open universe, you could travel in a straight line forever and ever and you never would get back to where you started.

The answer is likely to be found within the next decade, through a combination of more powerful Earth-based telescopes and the Hubble Space Telescope.

But the most important reason to put new observatories into space is this. Every time astronomers have opened up a new window in the sky, looking at some form of energy from space that had previously been inaccessible, we have always discovered new things. When Galileo turned his telescope on the sky, he discovered the moons of Jupiter, the phases of Venus, the mountains of the Moon, and proved that the Earth is not the center of the universe. Just in the space between the first and last drafts of this book, a whole new phenomenon was found: giant circular arcs in intergalactic space!

Knowledge from space will help unify all the laws of physics that determine the very atoms out of which we're made. By using new kinds of telescopes and searching parts of the electromagnetic spectrum never studied before, and by continuing to search the familiar parts with greater sensitivity than ever, we will surely find some of the missing pieces in the cosmic jigsaw puzzle.

THIRTEEN

Lasers, Antimatter, and Solar Sails

ADVANCED PROPULSION CONCEPTS

At JPL, engineer Ross Jones has a vision. He sees hundreds of grapefruit-sized spacecraft flying through the solar system on missions to planets, comets, and asteroids. Launched with technology being developed for the Star Wars defense program, and taking advantage of the remarkable developments in microminiaturization of the last two decades, we may be able to mass-produce tiny, relatively cheap space probes and shoot them into space inexpensively.

This is one of several radical innovations being explored in laboratories around the world, any one of which could revolutionize space travel: microspacecraft, ion drives, laser propulsion, solar sails and antimatter. Some of these may eventually replace conventional rockets; in the meantime, plans are afoot to beef up existing rockets.

Microspacecraft

Ross Jones was inspired by the space–age version of the cannon: the electromagnetic railgun. One type of Star Wars defense would require the launch of tremendous numbers of projectiles into space. To reduce the cost, researchers have been investigating variations on the technique that Jules Verne used to launch his fictitious astronauts in *From the Earth to the Moon* in 1865. Verne shot his voyagers from a giant cannon in Florida, not far from where the actual astronauts left a century later to go to the Moon.

This technology is not far off. Experimental railguns sit in laboratories now, and could allow the launch of tiny spacecraft from orbital bases in the next couple of years. In the Star Wars concept, railguns would probably be taken into space by rockets, and used there to launch payloads against enemy missiles. However, these same railguns could be used to shoot spacecraft into interplanetary space, and it's possible that they may ultimately be able to launch directly from the ground.

One research company, SPARTA, is doing research to see whether an existing 16-inch artillery gun could be used to shoot small payloads into orbit. Researchers at Maxwell Laboratories have developed electromagnetic guns that shoot a projectile that looks like a round of tank ammunition (not surprisingly, since this gun was originally developed for army tanks). By short-circuiting 4 million amperes of current, they create an intensely hot plasma that pushes the projectile out at 100,000 to 200,000 gees of acceleration.

The trouble with this approach is that it subjects the payload to huge accelerations. Astronauts would get squashed before they left the atmosphere. However, for electronics such forces may be tolerable. A ruggedly designed silicon chip, for example, isn't bothered by such accelerations. And while much scientific equipment would be too delicate for such treatment, it appears that there are useful instruments that could be made to withstand this type of launch. Even TV cameras and transmitters seem feasible. Microminiaturization has come so far that instrumentation that would once have required a large spacecraft can now be crammed into a grapefruit.

In one new breakthrough, the Massachusetts Institute of Technology (MIT) has built insectlike robots weighing a couple of pounds. An actual working six-legged walking miniature robot designed by Rodney Brooks of the MIT Artificial Intelligence Lab is about a foot long, has four microprocessors and six infrared sensors, and uses off-the-shelf model airplane control motors. Powered by three on–board batteries, it weighs 2 pounds. It's able to walk over rough terrain and follow a person's body heat—a robotic bloodhound!

(a)

(b)

(a) SPARTA's 16-inch gun, used to study the feasibility of launching small spacecraft for Star Wars.

(b) Pegasus, a 49-foot-long rocket designed by Orbital Sciences Corp. and Hercules Aerospace Co., that will be launched from under the wing of a B-52 bomber. It's a three-stage solid rocket that can put half a ton into low-Earth orbit. It will probably launch its first satellite in 1989.

Rodney Brooks's six-legged walking miniature robot.

Scientists are now starting to consider sending miniature robots into space. In the next decade or two, hundreds of these robots could descend on Mars, the asteroids, and elsewhere, scampering around on six legs like a benevolent plague of locusts. The proposed U.S./USSR Mars Sample Return mission could use half a dozen robots to roam to dangerous sites and return with bites of Mars to the mothership.

And this is just the tip of the microminiatization iceberg. Everyone has heard of silicon chips, the microelectronics that makes possible digital watches and personal computers. But what few people know is that similar breakthroughs have been made in the design of microminiature *machines* with moving parts. Tiny gears, levers, and other mechanisms too small for the naked eye to see have been made in laboratories, using technology similar to that used to make the microscopic wires and transistors of conventional silicon chips. And as with electronic chips, once you've figured out how to make one device, it is easy to make millions of them.

It was 1959, two years after *Sputnik*, when Jack Kilby of Texas Instruments invented the integrated circuit that would revolutionize our lives. Just one dozen years later, a team at Intel Corporation produced the first microprocessor, essentially a computer on a chip, using refinements of the integrated circuit. A decade from now, microscopic

> *Teams of nanomachines [molecular machines] in nature build whales, and seeds replicate machinery and organize atoms into vast structures of cellulose, building redwood trees. There is nothing too startling about growing a rocket engine in a specially prepared vat. Indeed, foresters given suitable assembler "seeds" could grow spaceships from soil, air, and sunlight.*
>
> *Assemblers will be able to make virtually anything from common materials without labor, replacing smoking factories with systems as clean as forests. They will transform technology and the economy at their roots, opening a new world of possibilities. They will indeed be engines of abundance.*
>
> —K. Eric Drexler, Engines of Creation

robots will probably be routinely manufactured, and they could change the way we explore the solar system.

At the first microspacecraft conference, it was suggested that such probes be sent to *hundreds* of asteroids. Pluto, the last unexplored planet, could be visited within a reasonable time by microspacecraft. Developments in microminiature TV were presented that allow microspacecraft to take pictures with about 2 pounds of gear!

The most bizarre and ingenious idea of all was a microspacecraft built like a maple seed, whose wings would allow it to spiral down through the atmosphere of a planet or of a satellite like Titan. Hundreds of maple-seed spacecraft could be sent all over Mars or Venus, taking pictures of the landscape as they flutter down. Conceived by James Burke of JPL, the concept has now been patented. (The Japanese should send inquiries to JPL.)

This is the type of technology that the Japanese are sure to develop if the U.S. does not. The Japanese are already seriously involved in space exploration, because they realize it has been a major stimulant to the high technology in which North America was once the undisputed leader. In addition to joint projects with the U.S., they have their own rockets and launch facilities. In some specialties, their microelectronics are second to none. They successfully launched their first two interplanetary spacecraft, destined for Halley's Comet, using their rocket base in southern Japan. We shouldn't be surprised to wake

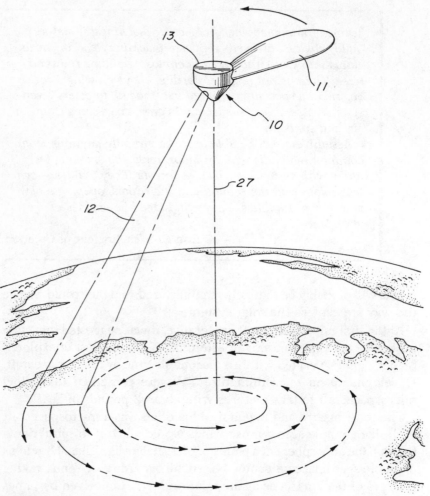

Patent drawing of James Burke's maple-seed spacecraft.

up one day to realize that the solar system is swarming with spacecraft marked Sony, Panasonic, and Toshiba.

And the Western Europeans are using space research as a tool to regain the eminence they once held in science and technology. As Western Europe approaches the economic unification of 1992, when borders will be eased, the European *Ariane* rockets continue to get bigger and to carry more of the world's commercial space traffic.

Launched in South America from French Guiana, *Ariane* could easily send microspacecraft on interplanetary missions.

Fortunately, the solar system is a big place. If every spacefaring nation starts sending out probes, we'll still barely penetrate what's out there.

Ion Drives

There's a wonderful technology that's been around for decades, yet has not been used: ion propulsion, using electrically charged particles accelerated to high speeds by electric fields. Ion engines produce a tiny thrust, but they can do it continuously, for years. The effect of such a thrust, day after day, is to accelerate the spacecraft to much greater speeds than is possible with practical chemical rockets.

Excellent ion thrusters already exist in the lab. All that is needed is the funds for a mission. The technology is already so well developed that NASA had planned to use it to send a spacecraft to Halley's Comet, until budget-slashers had their say.

A fascinating ion-drive project called Lunar GAS has been studied at JPL recently. It would be about the size of a barrel, small enough to qualify for the Shuttle's "Getaway Special" (GAS) rate that allows small payloads to fly at very cheap rates ($10,000 per flight). Covered with solar cells, the spacecraft could be dumped into orbit by the Shuttle, and its tiny ion drive would allow it to slowly spiral away from Earth until captured by the Moon's gravity. It would then spiral down into a low polar orbit, where its one instrument would search for water in the Moon's poles—one of the most important tasks for future space exploration.

This mission has not been funded, but it is so important and so inexpensive that I expect that this mission or something like it will fly in the 1990's.

Ion drives are excellent ways to get to other planets, where the long distances allow great speeds to be developed. In fact, JPL is even studying the design of what could be the first true near–interstellar spacecraft, *TAU*, described in the next chapter. Eventually, the technology could be adapted to piloted vehicles, so that human missions could get to Mars in months rather than years.

Solar Sailing

Another elegant idea NASA has explored (but which has never flown) is the solar sail, a way of taking advantage of the abundant solar energy in space. It uses the fact that light exerts pressure on any object. On Earth, that pressure is too feeble to be felt. It can be measured only by the most delicate experiments in the laboratory. But in space, if you put up a giant umbrella made of a very light material, such as Mylar or Kevlar, then sunlight exerts a gentle but noticeable pressure.

The Union Pour la Promotion de la Propulsion Photonique, known as U3P, is also seeking private sponsorship for its design and development of solar sailing spacecraft. U3P has persuaded Midi-Pyrenees, a regional government in France, to sponsor a race to the moon and to offer a prize. The custom of using races and contests to spur technology development has a long and respectable history, and so the idea of a solar sailing race to the moon has wide appeal. The project is especially exciting to earthbound sailors, who test their boats and prove their seamanship in races. The address for U3P is 6 Rue des Remparts, Cologny, Venerque, 31120, Portet sur Garonne.

The moon race does offer one problem, however. The sail is best operated in deep space over interplanetary distances, not near Earth where it must make many maneuvers. The race, then, will be quite demanding for a "first" sail; it will require sophisticated navigation, guidance, attitude control, dynamic and stability control, maneuverability, and preciseness. Nevertheless, the race is technically possible, and it should stimulate even more widespread interest in solar sailing.

The rules of the race are quite simple:
1. *All vehicles must start in geosynchronous orbit or lower. (Low Earth orbit would not be a good starting place, however, because of air drag and short maneuver time.)*
2. *All vehicles must use only photon pressure from sunlight on the Earth-to-moon transfer.*
3. *The first ship to go behind the moon wins.*

—Louis Friedman, Starsailing

As with the ion drive, the pressure is very small, but steady, so it can accelerate a spacecraft over a period of months with no rocket fuel at all. A private organization, the World Space Foundation in Pasadena, California, has been experimenting for years with solar-sail designs and hopes to launch the first solar sail in space in a couple of years. Solar sailing could prove to be the cheapest way yet of traveling through the solar system.

The similarities to sailing by boat are remarkable. Just as it is possible to travel upwind in a sailboat by tacking, it is possible to travel toward the Sun as well as away from it, by orienting the sail properly.

At first, this technology will be only used for robotic probes, but within 25 years, wayfarers may find themselves sailing very much as those aboard galleons once did—solar sailors sailing on sunbeams.

Laser Power

If the pressure of light from the Sun can push a sail gently around space, why not use an intense laser beam to do the same job? The force of light is not strong enough to compete with gravity, but by aiming a laser beam at a rocket, the propellant can be heated up far more than in a chemical engine, greatly increasing the thrust. Scientist Arthur Kantrowitz had this idea in 1972, and little work was done on it until recently. But now this looks as if it could be the breakthrough that makes human spaceflight almost as cheap as air travel.

Extremely powerful lasers have been developed, some for experimental plasma fusion reactors for energy generation and others to shoot down enemy missiles in the Star Wars program. Some of these lasers bear more resemblance to chemical rocket engines than to the weakly glowing red laser lights seen in bar-code readers in supermarkets. Chemical or electrical energy at the power source is converted into infrared light. If such a beam is aimed at a target, the pressure of light will push it, just like the sunbeam pushing on the solar sail. The thrust is increased if the beam vaporizes some material on the target, working like a rocket, except that the propellant on the target can get much hotter than rocket fuel, and the energy comes from the laser, not the fuel.

One approach would use space–based lasers to push spacecraft around, to transfer from one orbit to another. A JPL investigation

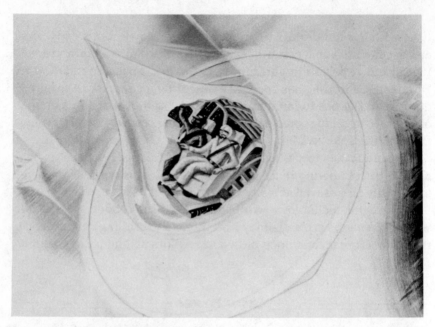

Artist Ray Rue's conception of the Apollo laser-propelled spacecraft designed by Leik Myrabo and colleagues.

showed that a 1-million-watt orbiting laser could be competitive with conventional rockets for powering an Orbital Transfer Vehicle, a tug to take spacecraft from the low-Earth orbit where the Shuttle flies to higher orbits where many satellites operate.

Theoretically, a spacecraft could be launched from Earth this way: A huge laser, with its great mass, would sit on the ground; the target would be mostly payload. Thus the power source itself does not have to accelerate, unlike a rocket, which has to carry great amounts of fuel with it. This would be vastly more efficient than a rocket could ever be.

At Lawrence Livermore National Laboratory, under the sponsorship of the Strategic Defense Initiative Organization, Jordin Kare and colleagues are studying laser propulsion. And engineering physicists Leik Myrabo, Jacques Richard, and colleagues at the Rensselaer Polytechnic Institute (RPI) have analyzed a Laser-Boosted Lightcraft Technology Demonstrator (LTD). This laser elevator could launch a 300-pound vehicle into Earth orbit, using a 100-million-watt laser array on the ground, consisting of a group of 1- to 3-million-watt lasers. The amazing fact is that most of this technology exists today. With funding, the Lawrence Livermore lab could launch such a spacecraft within five years!

The greatest virtue of this technology is that it would allow a tremendous volume of material to be launched into orbit at low cost. With a 20-million-watt laser on the ground, Kare estimates that several hundred *tons* of payload per year could be launched, using simple spacecraft, with a system costing far less than one Space Shuttle. The RPI group has designed more advanced missions, culminating in the Apollo Lightcraft that could launch a crew of five into orbit with a 2.5 billion-watt solar-powered space-based laser.

Kare goes so far as to suggest that missions to Mars might be possible via laser propulsion. He envisions the Mars vehicle and its fuel being lofted by laser into Earth orbit in chunks weighing a few hundred pounds. The crew and any hardware that couldn't be broken down would be carried up by the Shuttle or conventional rocket. If a laser launch facility had been set up on the Moon, that would be used, too, perhaps to provide oxygen or fuel extracted from lunar materials. The Mars vehicle would be assembled in orbit and blast off for Mars using conventional chemical fuels.

Orbital lasers could be used to shoot additional fuel and supplies to the astronauts while they're on their way. At Mars, they'd use conventional rockets to land and take off again. They'd return with the help of more laser-launched refueling vessels, returning the vessel to Earth orbit for eventual re-use.

Heavy-Lift Vehicles

At the opposite end of the launch-vehicle spectrum from microspacecraft are rockets being designed to be the next step beyond the Shuttle in terms of power. The Star Wars program needs to heft many tons of construction material into space, more than the present fleet of Shuttles could handle. This has led the military to study unpiloted rockets similar to the Saturn V that took us to the Moon. (Sadly, the Saturn assembly lines died with the *Apollo* program, when budget cutbacks forced NASA to put all their space eggs into the Shuttle basket.) Several heavy-lift vehicle options are being studied by the U.S. government:

Shuttle C.   This approach would use the Space Shuttle, but with an unmanned package replacing the orbiter that normally carries the crew and payload. On the existing Shuttle, the orbiter weighs about 130 tons, of which the payload to a 115-mile-high orbit is 24 tons. (The solid rockets and the huge external fuel tank bring the actual launch weight of the whole Shuttle to about 2,000 tons.) By replacing the crew with an automated system, much more of the 130 tons could be used for payload.

Shuttle II. This would be a redesigned Shuttle, probably with a similar payload, but cheaper to operate.

Titan V. The next step beyond the existing Titan IV rocket, which can launch 20 tons of payload to the same orbit as Shuttle C. The V would launch 30 to 75 tons.

Transition Vehicle. Another unpiloted rocket, this would be partially reusable, like the Shuttle.

Advanced Launch System (ALS). A completely new system being studied by the Air Force and NASA, it would probably be partially reusable: The booster would fly back to the ground, unlike the Shuttle's solid-rocket boosters that parachute to the ocean for recovery.

The National Aerospace Plane (NASP), also known as the Orient Express. Discussed in Chapter 2, this would be a piloted spacecraft that would take off like an airplane from a runway, breathe oxygen from the air so it didn't have to carry much of the heavy oxidizer, and fly into orbit.

The Russians have come to a similar conclusion about the need for heavy-lift vehicles, but have worked in the opposite way—from unpiloted to piloted. They recently flew the huge Energia ("energy") rocket, about as powerful as our old Saturn V. Energia is normally used as an unpiloted rocket for payloads of 100 tons, but they have adapted it for use with their space shuttle *Buran* ("snowstorm"), which is clearly

> *"It's not beyond reason to think that in 20 years we may have gotten to the point where somebody's figured out a clever way to allow people to take vacations up in space. One of these days Donald Trump will decide he wants to take a weekend vacation in space, and he can afford it."*
> —Edwin "Buzz" Aldrin, Apollo 11 *astronaut*

(a) NASA's proposed Shuttle C (Cargo), substituting a payload carrier for the familiar piloted orbiter. (Courtesy of Martin Marietta Corp.)

(b) A General Dynamics concept of an Advanced Launch Vehicle with a first-stage booster that flies back to the ground.

modeled on the U.S. Shuttle. This gives them a more flexible system than we now have, with the Energia available for raw power to put tonnage into orbit, and their shuttle available as an add-on when humans are needed.

The availability of heavy–lift launch vehicles will make it cheaper to build big structures such as space stations, solar power satellites (to beam power down to Earth), and, eventually, space settlements.

Antimatter

Antimatter sounds like science fiction, but is bread-and-butter reality to physicists. For decades, we have known that every elementary particle in nature has its opposite, just like the Chinese symbols of the yin and yang.

After electrons were discovered, anti-electrons (positrons) were found that had the same mass and spin as electrons but were positively charged instead of being negative. When an electron meets a positron, they annihilate each other, producing pure energy (in the form of gamma rays). This annihilation releases 100 percent of the energy that Einstein's equation, $E=mc^2$, says is contained within all matter. (By contrast, a hydrogen bomb only releases about 1 percent of the energy of the exploding matter. Fortunately, antimatter doesn't make good bombs—it fizzles rather than explodes.) Every elementary particle has its antiparticle: There are antiprotons, antineutrons, and so forth.

Antiparticles are produced in huge particle accelerators by physicists studying the structure of elementary particles like protons; so it's possible to manufacture antimatter. For example, atoms of antihydrogen

> Our Air Force is already funding research on antimatter, a far-out concept in which a positively charged proton and a negatively charged antiproton collide, annihilating each other and in the process releasing a burst of pure energy. These particles can be produced in minute quantities today, and could be the key to driving a star ship close to the speed of light.
>
> —Michael Collins, Liftoff

> *In the coming decades we will see the production and storage of significant quantities of mirror matter [antimatter]. The first uses for space travel will be within the solar system. And if no other propulsion system proves to be better, and we choose to spend the time to collect the solar energy to generate the kilograms of mirror matter needed, then another fantasy will become fact. One of these days we can ride to the stars on a jet of annihilated matter and mirror matter.*
> *—Robert L. Forward and Joel Davis,* Mirror Matter

can be built from positrons and antiprotons. Schemes have been developed to store the antimatter by suspending it within electromagnetic fields, to keep it from touching ordinary matter.

Scientist Robert L. Forward has been proposing antimatter propulsion for years at technical conferences and in books and an antimatter newsletter. While many scientists would say that antimatter will never be of practical use, since the amount of such matter in the world to-

> *When I think of the space technology of tomorrow, I think of three concrete images in particular. First, the Martian potato, a succulent plant that lives deep underground, its roots penetrating layers of subterranean ice while its shoots gather carbon dioxide and sunlight on the surface under the protection of a self-generated greenhouse. Second, the comet creeper, a warm-blooded vine which spreads like a weed over the surfaces of comets and keeps itself warm with super-insulating fur as soft as sable. Third, the space butterfly, a creature truly at home in space, acting as our agent in exploration and reconnaissance, carrying pollen and information from world to world just as terrestrial insects carry pollen from flower to flower. . . The Martian potato, the comet creeper and the space butterfly are merely symbols, intended, like the pictures in a medieval bestiary, to edify rather than to enlighten.*
> *—Freeman Dyson,* Infinite in All Directions

day is literally microscopic, the U.S. Air Force takes the concept seriously enough to fund some of this research, and the final stamp of approval came when the Rand think tank recently came out in favor of it.

We may be at a stage similar to Benjamin Franklin's time, when the
only source of artificial electricity was primitive batteries and crude
mechanical static-electricity generators. It would have been difficult
for Franklin to conceive that civilization two centuries away would
run on electricity.

Conclusion

Each of these technologies is highly promising. Each could result in
the drastic reduction of the cost of space exploration and commerce.
Star Wars, like it or not, is developing technology that can accelerate
our advancement into space: railguns, heavy-lift vehicles, super-
powerful lasers. Much of the technology of this century has been the
result of military research, from life-saving medicines to annihilation-
threatening nuclear weaponry. Now we can turn Star Wars swords
into spacefaring plowshares.

Almost certainly, at least one of these technologies will be a success
in the 1990's. Railguns, microspacecraft, solar sails, ion drives, and
heavy-lift vehicles will probably become reality. Small laser launches
could occur in the same period if supported financially. Antimatter
drives are further away from practicality but could become important in the early 2000's. The year 2000 should see a spacecraft pro-
liferation as dramatic as the rapid growth of aircraft in the mid-20th
century. The day when going into space as routinely as we now fly
from New York to London will be at hand.

FOURTEEN

How to Build a Starship

INTERSTELLAR TRAVEL

Pioneer 10 is, at this moment, truly living up to its name, flying into the edge of unexplored interstellar space, to be followed in the next few years by *Pioneer 11* and the two *Voyagers*.

Interstellar spaceships have long fascinated scientists and science-fiction writers, but there certainly will be no interstellar astronauts in the next 25 years. The distances are too vast and present technology too limited. However, there's a good chance there will be further robotic interstellar probes in this period. One of them, called *TAU*, is being seriously studied by JPL, and could be the first probe deliberately designed to explore interstellar space. The other, *Starwisp*, has been proposed by a remarkable scientist, Robert L. Forward, and is the logical candidate to be the second.

> *The quite reasonable conclusion is that by the middle of the next century we may well be able to afford to build starships.*
>
> Iain Nicolson, science writer

A Toe in the Cosmic Ocean: *TAU*

The distances to the stars are almost unimaginable. At the speed of *Pioneer 10*, it would take hundreds of centuries to reach the Sun's nearest stellar neighbor, the triple-star system Alpha Centauri. But at least we could start to explore the region of near-interstellar space beyond the planets.

At JPL, for example, engineers are doing the preliminary design of a spacecraft called *TAU*, which stands for Thousand Astronomical Units. (An astronomical unit is the distance of the Earth from the Sun: 93 million miles.) Under the leadership of the husband-and-wife team of Aden and Marjorie Meinel, it would explore space out to 1,000 astronomical units—about 30 times the distance of Pluto or Neptune, taking about half a century to do so.

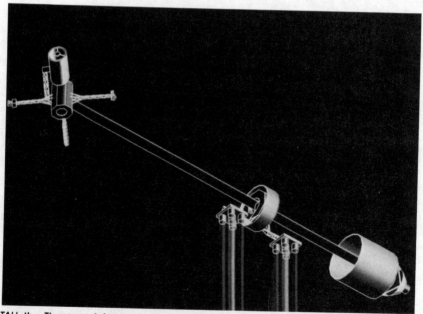

TAU, the Thousand Astronomical Unit mission, now being studied by NASA to be the first spacecraft designed as a galactic probe. It would use a nuclear power plant (the bulky object on one end) to run the ion-drive engines near the center. The other end contains a telescope to measure star positions.

*If the [matter-antimatter] engine is developed, a small,
highly specialized . . . vessel with a crew just large enough
to constitute a stable society could make the trip to Alpha
Centauri in only a few years. . . . The result would be a
slow but perhaps inexorable emigration to the planets of
other stars.*

—Robert Powers, Fellow of the Royal Astronomical Society.

*If science has strong support . . . and technology is being
intelligently used for the betterment of humankind, when
the time comes that we can build a starship, it will be built.*

Saul J. Adelman, astrophysicist,
and Benjamin Adelman, science writer

What makes it potentially practical is an ion-drive engine. An ion
drive uses electricity to accelerate ions (charged atoms) up to high
speed, as in the particle-beam weapons of the Star Wars program.
However, here the particles are the fuel. They shoot out the back like
a rocket, and the recoil pushes the spacecraft gently forward.

Ion drives generate too slight a push to take off from the Earth or
make it through the atmosphere, but the vacuum of space is ideal for
them. Unlike a rocket that shoots a powerful flame for a few minutes,
the ion drive shoots a tiny beam for months or years.

At first, the ion drive hardly moves at all. But if you let it continue
for months, it can build up enough speed to overtake a conventional
rocket. The tortoise wins out over the hare.

Ion drives need electrical power, and *TAU* would be going too far
from the Sun to use solar energy, so it would have to be powered by
a small nuclear generator of the sort that runs the *Pioneer* and *Voyager*
spacecraft. This would allow it to accelerate until it reaches a speed
of 200,000 mph, fast enough to get from the Earth to Pluto in an incredible year and a half, compared with the decade it would take for
Pioneer 10 to get there.

TAU would make possible precise measurements of the positions of the stars of our galaxy. One benefit of this will be to allow us to calculate the distance of many stars that are difficult or impossible to measure from Earth, by comparing their positions as seen from the spacecraft and from our planet. Using the surveyor's technique of triangulation, *TAU*'s path would serve as a vast baseline from which to compute their locations, and would probably detect some nearby stars currently lost among the millions of distant ones that crowd astronomers' photographs. We might even find stars closer to us than Alpha Centauri.

The Starship *Starwisp*

An even more ambitious project was designed by physicist Robert Forward, a man whose research has ranged from gravitational radiation to the use of antimatter. He works at the Hughes Aircraft Company Research Laboratories and has been in the forefront of research on

> It is difficult to go to the stars. They are far away, and the speed of light limits us to a slow crawl along the starlanes. Decades and centuries will pass before the stay-at-homes learn what the explorers have found. The energies required to launch a manned interstellar transport are enormous, for the mass to be accelerated is large and the cruise speed must be high. Yet even these energies are not out of the question once we move our technology out into space where the constantly flowing sunlight is a never-ending source of energy—over a kilowatt per square meter, a gigawatt per square kilometer. There are many ideas in the literature on methods for achieving interstellar transport. . . . In time, one or more of these idealistic dreams will be translated into a real starship.
>
> —Robert L. Forward,
> *Hughes Aircraft Company Research Laboratories*

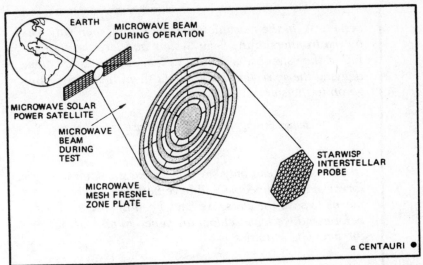

EARTH   MICROWAVE BEAM DURING OPERATION

MICROWAVE SOLAR POWER SATELLITE

MICROWAVE BEAM DURING TEST

MICROWAVE MESH FRESNEL ZONE PLATE

STARWISP INTERSTELLAR PROBE

α CENTAURI ●

Starwisp: Robert Forward's proposed tiny robotic interstellar space probe we may be able to build in the near future. It would use a proposed solar-powered satellite (designed to beam power to Earth). For a brief time, the satellite would beam microwave energy through a "lens" to a very light-weight sail, pushing it until it reached one-fifth the speed of light.

possible robotic—and even human—interstellar spaceships; it should not be surprising that he is also a successful science-fiction novelist.

He has devised on paper an ingenious robotic spacecraft he calls *Starwisp,* one that might be the first we send to another star. Even the ambitious *TAU* project would travel a distance of less than a percent of that to Alpha Centauri. *Starwisp* could go all the way to the star system and still return data to us.

The secret is in ultraminiaturization. Forward proposes a spacecraft weighing about an *ounce*, consisting mainly of metal mesh that would be pushed by microwaves beamed from the Earth. He would like to use a solar-powered satellite of the sort that physicist Gerard O'Neill proposes to build in order to beam solar energy to Earth. Forward would simply divert the microwave beam of that power station for a few minutes, and use that energy to push the mesh of *Starwisp*, accelerating it until it reaches 20 percent of the speed of light.

After that, *Starwisp* would travel silently across the light-years un-

Sometime, in the not unforeseeable future, men will stand
on the frontiers of the Solar System and face the stars.
Either they step back, content to live out the future alone,
denying the spirit that has brought them thus far—or they
go on to the stars. Which is it to be?

James Strong, Fellow of the British Interplanetary Society

In the end, we can only say this: If space colonization goes
forward, in time people will have both the ability and the
means to seek the stars. We have been in such situations
before, and we have seldom disdained to take advantage of
the new opportunities.

T.A. Heppenheimer, aerospace engineer

til it reached the star 21 years later. Then, a power station would again
be pointed at *Starwisp*, flooding the star system with faint microwaves
that would power the electronic circuits at the junctures of the mesh.
Light detectors would form images of the planets and other sights of
that system and radio them back, using the mesh as a large antenna
to beam the signals to the Earth. Several years later, we would see the
first TV pictures from another star system.

This is something that could almost be built with today's technology,
and could fly within 25 years.

Forward has also been a leader in the design of human spaceships
for interstellar flight, although we are not likely to see these for at least
a century. The significance of his designs, and those of other scientists who have also studied the problem, is that we can see right now,
with mere 20th-century technology, ways that it is theoretically possible
to send people to other stars.

All the schemes for piloted interstellar spaceships require vast
amounts of energy, but if we build solar-powered satellites, or if we
harness the energy of the Sun in thermonuclear fusion power plants,
as several nations are now trying, then the amount of energy available

from our civilization might enable us to send a few Earth representatives on interstellar jaunts.

That we can even conceive of methods to get between the stars means that it is possible, and once something is possible, it usually gets done, though sometimes by methods we never anticipated. The advances of science and technology in the next century might therefore result in currently inconceivable ways of traveling between the stars—warping space, perhaps, as in *Star Trek*.

Only 66 years elapsed between the Wright brothers' first flight and the Apollo landing on the Moon. I am confident that, if civilization endures, some of our descendants are going to walk on the planets of other stars, and probably sooner than most people think.

FIFTEEN

In Search of E.T.

THE SEARCH FOR EXTRATERRESTRIAL INTELLIGENCE

There are projects operating at the present moment that sound like science fiction, but are really science fact. The endeavor is called the Search for Extraterrestrial Intelligence (SETI), and its disciples have a very good chance of succeeding within the next two decades.

> Heaven and earth are large, yet in the whole of space they are but as a small grain of rice. . . . It is as if the whole of empty space were a tree, and heaven and earth were one of its fruits. Empty space is like a kingdom, and heaven and earth no more than a single individual person in that kingdom. Upon one tree there are many fruits, and in one kingdom many people. How unreasonable it would be to suppose that besides the heaven and earth which we can see there are no other heavens and no other earths?
>
> —Teng Mu, 13th-century philosopher
> (translated by Joseph Needham)

SETI (rhymes with "jetty") projects are being built and operated around the world. Ohio State University has been operating such a project for a decade; the world's most advanced operational SETI system is at Harvard; and NASA is building an even more powerful setup that should be fully operational in the late 1980's. The Soviets have SETI projects of their own, and will be conducting some of their search from space.

E.T. Comes Out of the Closet

What brought SETI into the world of respectability was, in my opinion, the launch of *Sputnik* in 1957. Overnight, Buck Rogers fantasy became Walter Cronkite reality.

It was no coincidence that within just two years, the first modern scientific paper proposing SETI appeared in the British journal *Nature*. *Nature* is more accustomed to publishing the research of Nobel-Prize-winning scientists than such far-out speculations. Indeed, the first paper by Watson and Crick describing the double-helix structure of DNA appeared in that journal. But *Sputnik* created an atmosphere in which scientists could seriously discuss the possibility of other civilizations without being subjected to ridicule.

This pioneering paper, by scientists Giuseppe Cocconi and Philip Morrison, proposed that the easiest way to detect another civilization would be by radio signals. Radio travels at the speed of light, and a large band of frequencies (microwaves) can travel throughout the entire Milky Way with little interference. Already, we can broadcast to nearby stars using off-the-shelf technology. Interstellar spaceships may one day be built, but they clearly require a far more advanced technology and would be considerably more expensive than radio transmitters and antennas.

There may well be methods for communicating by using techniques unknown to today's science, but if an alien society wants to communicate with *anyone*, not just the most advanced beings, it will have to use a technology known to all civilizations, and radio is probably the most primitive means for the job. It's the "Boy Scout" technology of the universe.

Thus, in 1960, astronomer Frank Drake conducted the first modern search for artificial radio signals from another civilization. He called it Project Ozma, after the queen of the Land of Oz. He began a search that has been continued off and on by scientists all over the world.

Despite the gift of intelligence, the gift of mobility, the gift of historical perception, the gift of anticipation, human beings are preoccupied with undertakings that can make life on Earth uninhabitable. Nothing we make on Earth is in greater abundance than destructive force. We have amassed 30,000 pounds of destructive force for every man, woman, and child on Earth. We don't have 30,000 pounds of food in reserve for every human being on Earth, or 30,000 pounds of medicines, books, or any of the things that ennoble life. But we have an infinity of force to use against one another. In the middle of a forest of bombs on Earth, it is difficult to see the tree of life.

Bertrand Russell once said that man can never resist any folly of which the human mind is capable. It is quite possible that the folly we have known on Earth has existed elsewhere in the universe. It is quite possible, however, that there are answers, better answers, than we have been able to find to our problems and our delusions. Ultimately, I think the question that must ignite the human mind in connection with the Viking trip to Mars has to do with our loneliness in the universe. We are transported by the notion that there may be other humans out there too. It is almost unscientific to think that life does not exist elsewhere in the universe. Nature shuns one of a kind. Infinity converts that which is possible into the inevitable. The fact that we are attempting to find out where and how may be the answer to the question, "why explore the universe?"

—Norman Cousins, author

The Cosmic Haystack

The basic problem with SETI is that the sky is big and the number of frequencies to be searched immense. It's like looking for a needle in a haystack, and is thus often called the problem of the Cosmic Haystack. Fortunately, scientists keep coming up with schemes to search more and more channels simultaneously, or to guess intelligently which frequencies an alien civilization might broadcast on.

Three SETI projects currently stand out in the U.S. One is the world's

This fact lies behind a remarkable event in human history. Almost imperceptibly, without really intending it, within the past two or three decades we have entered a new communicative epoch. Until that time, we could have made no sound, no pattern or mark, no explosive flash of light on our small planet that could be detected far out among the stars by any means we understand. Space is too deep, and the stars are rivals too brilliant, for any mere faint human glow to become visible far away. Even the whole amount of sunlight reflected from a planet, a light source thousands of times more powerful than all the energy now at human disposal—is still beyond our ability to pick out at the distance of a nearby star. But our radio technique, only a generation or so old, has now reached such maturity that a signal sent from an existing radio dish on Earth, with sending and receiving devices already at hand, could be detected with ease across the Galaxy by a similar dish, if only it is pointed in the right direction at the right time, tuned to the right frequency. Such a lucky observer—or one who is patiently and systematically searching—would see us as unique, distinguished among all the stars, a strange source of coherent radio emission unprecedented in the Galaxy.

Or are we without precedent? Are we the first and only?

Or are there in fact somewhere among the hundred billion stars of the Galaxy other such beams, perhaps so many of them that our civilization, like our Sun, is to be counted as but one member of a numerous natural class? For such a radio beam cannot come, we think, from any glowing sphere of gas or drifting beam of particles. It can come only from something like our own complex artificial apparatus, far different from any star or planet, smaller, newer, much more particular; something we would recognize as the product of other understanding and ingenious beings. . . .

We do not intend to send any signals out to add to those which have already gone out from our TV transmitters and our powerful radars. Rather, we want to listen, to search all the directions of space, the many channels of the radio (and other) domains, to seek possible signals. Perhaps it will be only an accidental signal, as we have made ourselves. That would be harder to find. Or perhaps there is a deliberate signal, a beacon for identification, or even a network of communication. There seems no way to know without trying the search. This is an exploration of a new kind, an exploration we think both as uncertain and as full of meaning as any that human beings have ever undertaken.

The search would be an expression of man's natural exploratory drive. The time is at hand when we can begin it in earnest. How far and hard we will need to look before we find a signal, or before we become at last convinced that our nature is rare in the Universe, we cannot now know.

—*Consensus of the Science Workshops on Interstellar Communication,* The Search for Extraterrestrial Intelligence, *NASA Special Publication 419*

longest-running SETI program. At Ohio State University, where hardworking John Kraus and Robert Dixon have run their project on a shoestring, this search has been operating for about a decade. Using all the equipment they could beg, borrow, or get donated, they have struggled against great odds, the greatest of which was a Close Encounter of the Real Estate kind.

The developer wanted to extend a golf course onto land that their radiotelescope antenna occupied. He had bought the land and was set to start work on it, but an international uproar developed over the idea that Arnold Palmer was considered more important than E.T., and the developer finally leased the land to the intrepid researchers.

The Ohio State project continues, with difficulty, to this day.

NASA is the obvious organization to sponsor SETI, but their attempts have met with obstacles as troublesome as a golf course. In 1981, NASA proposed building a large-scale, fully computerized SETI project, but it was killed in Congress by Senator William Proxmire of Wisconsin, who thought the taxpayers' money shouldn't be spent on a search for little green men.

In the wake of a congressional budget amendment (instigated by Proxmire) that explicitly prevented NASA from funding SETI, the only way for American scientists to pursue SETI was for non-NASA people to use private funds. But where could such people be found, and where would the funds come from?

The first answer turned out to be a Harvard professor, Paul Horowitz, who was visiting NASA's Ames Research Center in northern California, where he studied the proposed NASA design. He conceived the possibility of making a small-scale, special-purpose, computer-controlled receiver combining his own ideas with those of NASA. He proposed concentrating on "magic" frequencies, bands that nature broadcasts on, such as that of hydrogen, 1420 MHz. Because his proposed project was much cheaper than the $1.5 million-dollar-a-year NASA proposal, it could be funded by a private organization, thus bypassing the congressional obstacle course.

But who could come up with the $10,000 needed to start building the hardware?

The Arecibo dish in Puerto Rico. It is the world's largest radiotelescope antenna, 1,000 feet in diameter, and has been used occasionally to search for signals from other civilizations.

Fortunately, in 1979, a private organization had formed that was devoted to the exploration of the solar system and the search for extraterrestrial intelligence. Called The Planetary Society, it was the joint conception of Cornell astronomer Carl Sagan; Bruce Murray, then head of JPL; and Louis Friedman, then a JPL scientist.

In the grim atmosphere following the congressional amendment, the new society studied Horowitz's proposal and decided that a low-cost but effective SETI project was an excellent way to fill in temporarily for NASA. They agreed to fund Horowitz.

He immediately set to work, and built and tested his system within the next year. It was so portable that he called it "Suitcase SETI"; the idea was to take it from one radio observatory to another whenever time was made available for its use.

Radio astronomy, however, is a busy profession. Radiotelescopes are powerful tools, in demand to monitor the universe. Mother Nature has filled our sky with so many kinds of mysterious objects emitting radio signals that there are always long waiting-lists for the world's major radiotelescopes. There's never enough time for all the astronomers to make all the observations they want.

Pity the poor SETI researcher: Even after he has survived the attacks of congressmen and the headaches of state-of-the-art electronics, he has to jump into the same pool as all the other radio astronomers who want to scan the sky.

But SETI is a project that cannot guarantee results. We'll have to search for years to have a reasonable chance of success. Thus the scientists who schedule radiotelescopes, who live and die by the motto "publish or perish," are reluctant to make the instruments available for something as iffy as SETI.

Imagine Horowitz's emotions when he took Suitcase SETI back to Harvard and found that they had an old 84-foot-diameter radiotelescope dish antenna they were about to mothball! Quickly, he asked Harvard if he could have it, provided he could find funds with which to refurbish and operate it, and then he asked The Planetary Society for the money. Both institutions agreed, and Suitcase SETI was installed there as Project Sentinel. In 1983, Sentinel was turned on and began scanning the sky on 131,000 radio channels.

The next figure who altered the course of SETI history was E.T.'s friend, Steven Spielberg. Spielberg, the man who gave us the blockbuster hits *Close Encounters of the Third Kind* and *E.T.*, decided to donate $100,000 to The Planetary Society to upgrade Project Sentinel. The man who did more than almost anyone else to popularize the concept of *nice* extraterrestrials, after Hollywood had given us years of nasty ones, now made it possible to expand the system to an unprecedented eight *million* channels. *E.T.* was paying for the search for E.T.

The newly expanded Harvard facility is now running under the name Project META, for Megachannel Extraterrestrial Assay. Project Sentinel was already the most advanced SETI system currently operating on Earth, but Project META is a quantum leap ahead of that.

While Suitcase SETI was being built, NASA tried to resurrect the

One of the requirements imposed on junior members of the galactic community might be to establish beacons to attract and eventually to educate the developing intelligent races. If all this be true, a vast body of knowledge accumulated over billions of years awaits access by mankind, by any race that has technological prowess to qualify for it. This body of knowledge might be termed our galactic heritage. Access to the galactic heritage would truly be the most important event in the recorded history of man.

Benefits would be threefold—scientific, cultural, and societal. We would expect to find answers to questions that now puzzle scientists, and answers to questions not yet even asked. Whole new natural sciences may be waiting to be discovered. We would also know how the universe actually evolved and whether it is slowing to an ultimate death or headed for a phoenix-like recycling. We would learn of new art forms and aesthetic experiences. But perhaps most important, we might learn how to live in harmony with ourselves.

The galactic community must already have distilled out of its member cultures the political systems, the social forms and the morality most conducive to long-term survival— survival not just for a few generations but until each star dies. We might learn how other races solved their problems and how they took their responsibility for genetic evolution in a compassionate society. If [SETI] could provide even a few of these answers, its cost would be justified many times over.

> *Participation in a galactic community would, in astronaut Neil Armstrong's words, "enhance the spirit of man." It would lift our horizons out of the sphere of our own petty rivalries and involve us in the common cause of life throughout the Galaxy. Our rendezvous with destiny may lie in this contact with other intelligent life. Some day, cosmic life might save the whole universe from its expansion or contraction into oblivion, just as life on Earth saved our planet from the heat death experienced by Venus. The end of mankind's childhood could very well be discovered at the cosmic "water hole." It is up to us to decide whether or not man should now begin an effort to end his isolation in the universe.*
>
> *—Bernard Oliver, NASA Ames Research Center*

SETI proposal killed by Congress. When scientists such as Carl Sagan convinced Senator Proxmire of the worth of the project, he withdrew his objection, and the next year it was funded. It is now being built and should start full-scale observations around 1992 if Congress continues its support.

It will have about the same number of channels as META, but will be even more powerful, covering not just small portions of the radio spectrum near magic frequencies but essentially the entire microwave radio spectrum. Part of the system, designed by engineers at NASA's Ames Research Center and Stanford University, will look at nearby Sun-like stars. The other part, JPL's, will search the whole sky, though at reduced sensitivity.

The Russians, who have conducted several searches of their own, are building a radiotelescope near Samarkand to be used part-time for SETI, and one near Gorky to be devoted entirely to SETI. Recently, another Soviet group began an optical search for light-signals from other civilizations. In addition, the Russians plan to launch a radiotelescope into orbit that may spend some of its time on SETI.

Canada, France, West Germany, Holland, and Australia have all conducted brief SETI searches, and Japanese scientists have proposed using their new 150-foot-diameter radiotelescope, one of the best in the world, located at Nobeyama, Japan.

NASA and The Planetary Society hope to conduct SETI from the Southern Hemisphere in addition to the U.S. If the alien civilization with the loudest transmitters should happen to be visible only from the Southern Hemisphere, we'll need to do a lot more work down there than the few Australian searches that have been run. Argentina has two radiotelescopes that could be used, and the Argentines have recently started a SETI project. In addition to their own SETI work, the Argentines recently concluded an agreement with The Planetary Society to begin building a duplicate of the Harvard META system that could begin observations as early as 1990.

What If . . . ?

The sky is now being scanned in many ways by many projects. Most of what we've learned about the universe in the last decade seems to increase the likelihood that life exists elsewhere. If so, some of it has probably evolved intelligence. In my opinion, the chances are excellent that, well within the next 25 years, someone will find the first indisputable proof of an extraterrestrial civilization.

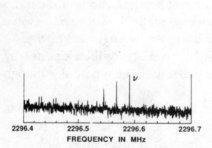

RECEIVED SPECTRUM
WHILE TRACKING VOYAGER 2

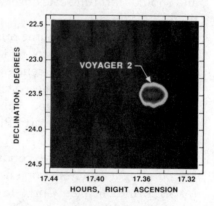

SKY SURVEY MAP
VICINITY OF VOYAGER 2
(AT FREQUENCY ν)

A test of the NASA SETI system. Using just a small antenna, they were able to detect the *Voyager 2* spacecraft about two billion miles away. Our first detection of another civilization might resemble this test: spikes at left show the radio signal sticking up above the hiss of the cosmos; at right, a map of the sky shows a "hot spot" where the spacecraft is.

THE IMPACT OF SETI

Whether the search for extraterrestrial intelligence succeeds or fails, its consequences will be extraordinary. . . .

Some have worried that a message from an advanced society might make us lose faith in our own, might deprive us of the initiative to make new discoveries if it seems that there are others who have made those discoveries already, or might have other negative consequences. But we point out that we are free to ignore an interstellar message if we find it offensive. Few of us have rejected schools because teachers and textbooks exhibit learning of which we were so far ignorant. If we receive a message, we are under no obligation to reply. If we do not choose to respond, there is no way for the transmitting civilization to determine that its message was received and understood on the tiny distant planet Earth. (Even a sweet siren song would be little risk, for we are bound by bonds of distance and time much more securely than was Ulysses tied to the mast.) The receipt and translation of a radio message from the depths of space seems to pose few dangers to mankind; instead it holds promise of philosophical and perhaps practical benefits for all of humanity.

Other imaginative and enthusiastic speculators foresee big technological gains, hints and leads of extraordinary value. They imagine too all sorts of scientific results, ranging from a valid picture of the past and the future of the Universe through theories of the fundamental particles to whole new biologies. Some conjecture that we might hear from near-immortals the views of distant and venerable thinkers on the deepest values of conscious beings and their societies. Perhaps we will forever become linked with a chain of rich cultures, a vast galactic network. Who can say?

If it is true that such signals might give us, so to speak, a view of one future for human history, they would take on even greater importance. Judging that importance lies quite outside the competence of the members of this committee, chosen mainly from natural scientists and engineers. We sought some advice from a group of persons trained in history and the evolution of culture, but it is plain that such broad issues of the human future go beyond what any small

> committee can usefully outline in a few days. The question
> deserves rather the serious and prolonged attention of many
> professionals from a wide range of disciplines—anthropolo-
> gists, artists, lawyers, politicians, philosophers, theologians—
> even more than that, the concern of all thoughtful persons,
> whether specialists or not. We must, all of us, consider the
> outcome of the search. That search, we believe, is feasible;
> its outcome is truly important, either way. Dare we begin?
> For us who write here that question has step by step
> become instead: Dare we delay?
>
> —*Consensus of the Science Workshops on Interstellar*
> *Communication*, The Search for Extraterrestrial Intelligence,
> *NASA Special Publication 419*

The proof may come from one of the SETI projects.

Or it might be an accidental by-product of conventional astronomy. Perhaps somewhere in the mountains of computer tapes from the *IRAS* satellite, there is the first sign of a Dyson Sphere, the hypothetical result of an advanced civilization building a complete shell around its stellar system.

Or perhaps the Hubble Space Telescope will find some unnatural alignment of stars.

Or one of our space probes may find some alien artifact.

Or perhaps the electronic-warfare receivers of the American National Security Agency or the Soviet KGB will pick up something even more mystifying than they're used to deciphering.

Or maybe some computer hacker with a satellite dish and a home-brew receiver will find strange signals not of this world.

Sooner or later, I suspect, we will detect another intelligent species.

The simple knowledge that we are not alone will affect our thinking even more radically than did *Sputnik* and the later sights of the Earth from space. Suddenly, it will be hard to take so seriously the differences between Americans and Russians or Arabs and Jews, when compared to the Little Green Men.

And if that happens, we on this planet will turn all our greatest telescopes onto that point in the sky beckoning to us. We may find television signals showing pictures of life on other worlds. We may find a dictionary to teach us how to speak Galactic. We may find an *Encyclopedia Galactica* with the history culture, music, religion, and science of a society billions of years older than ours.

We may find that the solutions to the problems of overpopulation, pollution, resources, disease, and war were found long ago, and the answers are streaming toward us at this very moment.

SIXTEEN

Footsteps into the Universe

THE FUTURE

Space is only a hundred miles over our heads, about the distance between New York City and Philadelphia. Yet it has taken us millions of years to cross that distance. In the next century, I expect, it will be as routine to travel there as it is today to shuttle back and forth between two cities.

Space is, as *Star Trek*'s Captain Kirk said, the final frontier. It is where the answers to many of our problems lie. It is where our industry ought to be focused, to free this planet to be the paradise it should be.

It is a place of unlimited resources, which we can harvest without harming the Earth's environment. It is a place where medicines and new materials can be manufactured. History shows the steady flow of humanity to habitable, commercially useful places and the incessant enlargement of cities, strongly suggesting that space is a place where many of us will live in the 21st century, some of us for a few

There are two futures, the future of desire and the future of fate, and man's reason has never learnt to separate them.

—*J.D. Bernal, scientist*

Artist Jon Lomberg's vision of a space colony, a place where thousands of people could live in comfort.

days, perhaps in hotels orbiting the Earth, and others for whole lifetimes, working there. It is a place where the answers to some of our most profound questions about the nature of life and the nature of the universe may be found.

People often think the U.S. is spending far more on space than it is. The budget of NASA is now about $10 billion per year, including the cost of replacing the *Challenger*. Compare this figure to a few other spending figures for 1986: $18 billion on international affairs, $269 billion on social security and medicare, or $286 billion on defense. About one cent in every tax dollar goes to the civilian space program.

The Near Future

For the near term, we have to make some important decisions about the future of the American space program. Since the *Apollo* program, NASA has had few long-term goals. Scientists, futurologists, representatives of ordinary citizens, and members of industry need to come together to form specific targets for the program.

We need to decide how much the country can afford to spend on space, bearing in mind that research has almost always more than paid for itself in its returns to society. Since the piloted space program seems popular enough to continue funding, we need to be sure that a healthy program of space probes is not lost in the budget debates over the Shuttle program and Space Station. We should try to keep the unpiloted program at a steady level, instead of letting it rise and fall so greatly that it drives skilled people away from the NASA centers.

We need to decide how much will be done at the taxpayers' expense and how much by private citizens, through industry. There has been a tendency until recently for NASA to try to do almost everything,

We Americans suffer from too much data, too many facts, at times. We are bombarded by it on our television. One of the problems we've had the last few years, that NASA has had, is that we have seen almost too much space and have seen the wrong kind. We have been given the facts over and over again, and they are always diminished by what I call the aesthetic of size. Television diminishes everything it touches and makes it small. It takes a rocket that is 300 feet high and crushes it down to a 14-inch image.

—Ray Bradbury, author

and it's healthy to transfer economically sound parts of the program—such as communications satellites—to private industry, where competition will keep costs down. And we need to eliminate governmental obstacles to private space programs.

Many of the answers depend heavily on what we really want to do in the long run. The design of the Space Station will be determined by whether we want to use it for industry, or for scientific research, or for preparing for a lunar outpost, or for an asteroid mission, or for a human mission to Mars.

We need to involve our international colleagues more in these decisions, since such collaboration can greatly reduce the cost to the American taxpayer. The Japanese and the Western Europeans are determined to be involved in space research, and are willing to spend billions to help us build the Space Station.

Those two factions have also developed their own rocket-launch facilities in Japan and in South America. The Chinese and Russians, of course, have their own, too; Brazil is also starting to launch orbital

A possible nuclear-powered ion-drive mission to explore Saturn's rings.

rockets. This worldwide growth in space capability guarantees that competition will keep costs coming down, and soon no one disaster will be able to slow down space development as drastically as the *Challenger* did.

The Soviets have indicated their willingness to cooperate on certain projects, possibly including a joint human mission to Mars. We must take advantage of this opportunity to combine Western and Soviet strengths, reducing the cost of scientific exploration for all.

Individual citizens can help make these decisions by writing to their congressmen. So few letters are received on most subjects that a letter supporting a particular space project, or space exploration in general, will have much more impact on a congressman than you might expect.

The disaster of the *Challenger* explosion has given us an unusual opportunity to make decisions in a period of calm.

The U.S. needs to have a steady, long-term plan for space exploration and use, though it must be flexible enough to allow for scientific discoveries that change the direction of research, and for inventions that change the way we build our space vehicles. Instead of jumping this way and that, depending on our disasters or Russia's successes or the political mood of the moment, we should have a program to explore the solar system systematically with robotic probes, revisiting the most interesting places that we found during the first generation of space exploration. To me, the most important robotic projects not yet funded are: a lunar polar orbiter to detect water on the Moon; a Mars rover to search for evidence of past or present life there and to pave the way for a human expedition; and a Titan probe to explore that extraordinary moon.

We should not neglect conventional astronomy, and should build Earth- and space-based observatories to systematically cover all sources of information about the universe that we are aware of: the electromagnetic spectrum, cosmic rays, neutrinos, and gravitational waves.

We need to have some practically oriented research to lay the foundations for a space industry that can gradually be taken over by private companies without further taxpayer expense, returning to the economy the benefits of space in much the way that research on microelectronics has created employment and profits for many.

We need to decide our next major astronaut missions. Shall we go back to the Moon, or to an asteroid, or to Mars? At this time, we can-

RATIONALE FOR EXPLORING AND SETTLING THE SOLAR SYSTEM

The Solar System is our extended home. Five centuries after Columbus opened access to "The New World" we can initiate the settlement of worlds beyond our planet of birth. The promise of virgin lands and the opportunity to live in freedom brought our ancestors to the shores of North America. Now space technology has freed humankind to move outward from Earth as a species destined to expand to other worlds. . . .

The settlement of North America and other continents was a prelude to humanity's greatest challenge: the space frontier. As we develop new lands of opportunity for ourselves and our descendants, we must carry with us the guarantees expressed in our Bill of Rights: to think, communicate, and live in freedom. We must stimulate individual initiative and free enterprise in space. . . .

Historically, wealth has been created when the power of the human intellect combined abundant energy with rich material resources. Now America can create new wealth on the space frontier to benefit the entire human community by combining the energy of the Sun with materials left in space during the formation of the Solar System. . . .

In undertaking this great venture we must plan logically and build wisely. Each new step must be justified on its own merits and make possible additional steps. American investments on the space frontier should be sustained at a small but steady fraction of our national budget. . . .

In his essay Common Sense, *published in January of 1776, Tom Paine said of American independence, "Tis not the affair of a City, County, a Province, or a Kingdom; but of a Continent. . . . Tis not the concern of a day, a year, or an age; posterity are virtually involved in the contest, and will be more or less affected even to the end of time, by the proceedings now." Exploring the Universe is neither one nation's issue, nor relevant only to our time. Accordingly, America must work with other nations in a manner consistent with our Constitution, national security, and international agreements. . . .*

> As formerly on the western frontier, now similarly on the
> space frontier, Government should support exploration and
> science, advance critical technologies, and provide the
> transportation systems and administration required to open
> broad access to new lands. The investment will again
> generate in value many times its cost to the benefit of
> all. . . .
>
> When the first Apollo astronauts stepped onto the Moon,
> they emplaced a plaque upon which were inscribed the
> words, "We came in peace for all mankind." As we move
> outward into the Solar System, we must remain true to our
> values as Americans: To go forward peacefully and to
> respect the integrity of planetary bodies and alien life
> forms, with equality of opportunity for all.
>
> —Pioneering the Space Frontier,
> *The Report of the National Commission on Space, 1986*

not afford to do all three, so we should choose one that excites the most interest among our potential collaborators, to reduce the costs and increase international cooperation.

We need to get rolling on the next generation of spacecraft, the aerospace plane, which could greatly reduce the cost of piloted missions, and set the stage for permanent habitation in orbit; perhaps this plane will even make space tourism possible.

I, personally, think we should develop a permanent human presence in space in three stages: first, by building a modest, flexible, permanent Space Station where it will be possible to perform materials research and to study the long-term effects of weightlessness on humans. We could use that as a way station for the second stage: A permanent lunar base similar to our Antarctic bases. There, we could begin to learn to harvest the resources of space and to solve the problems of living on another world permanently, while at the same time doing good scientific research. This would give us the experience of low gravity and closed ecologies that would serve as the foundation for the third step: a human expedition to Mars. The use of lunar resources could reduce the costs of building the Mars ships.

And building Jerome Wiesner's international Earth-watching station will go a long way toward ensuring that there's a civilization for our astronauts to return to.

New Space Drives

New technology will revolutionize space travel.

Modified artillery guns or electromagnetic railguns will shoot small, tough payloads into space. This will drastically reduce the cost of resupplying orbiting astronauts, and will go far to make space stations and colonies economical.

Microelectronics shows no sign of slowing its explosive (or implosive) growth, and micromachinery looks like it's going to follow the same path. And the power of computer software continues to increase dramatically, making those small packages ever smarter.

Ion drives and solar sails will fly soon. Ion drives can be powered by solar cells, so both types of drive will harness sunlight for cheap space travel. Eventually, these will probably be adapted for human spaceflight, making it easier to colonize places like Mars.

With these technologies, in a decade we should see hundreds of tiny spacecraft probing every nook and cranny of the solar system.

The development of powerful lasers for fusion and for Star Wars will make launch of small payloads by light-beam practical very soon. Within five years, we could see the first small payload lofted into orbit, the *Sputnik* of the Laser Age. If this proves as attractive as it now looks, humans could be traveling into space on beams of light before the 1990's are over. One of the beauties of this is that you don't need to lift huge spacecraft into orbit. Most of the weight stays on the ground, and even if all you do is carry one person at a time, you can do it endlessly, as many times as you want. Lasers may make the next century the age of the space elevator.

And one other new concept now being explored by NASA and industry is *tethers*, lines miles in length that can drop probes into the atmosphere, exchange energy between two spacecraft, or transmit power from a large vessel to a small probe. By making conductive tethers, we could generate power directly from Earth's magnetic field.

The most important lesson is that industry and government should invest in each of these alternatives. Only in this way can we be sure that we don't miss the best bet. It would be unwise to rely on just one type of system, as we've learned with the Shuttle. We will probably find the history of transportation repeating: No one type of ship was ever ideal for all kinds of water travel. Sailing vessels, steamboats, and diesel ships all had their places. The same will be true of space vehicles.

We should be able to drastically reduce the cost of getting a pound of payload into space. I expect that by the year 2000, you will be able to drop a package off at Federal Express and have it delivered into orbit for not much more than the cost of air freight to Japan. And soon after, space hotels should become practical.

The Long Run: Space Colonization

In the long run, the solar system will be our playground and our living room.

Some years ago, Princeton physicist Gerard O'Neill posed a problem to his students: Just for fun, could they solve some of the problems involved in building a space colony orbiting the Earth? He intended it only as a teaching trick to make the class more interesting, but to his amazement, students came up with potentially practical solutions to all the problems he could pose in the design of a space colony. He suddenly realized that such colonies were not something out of future centuries of history but could be built right now.

Furthermore, he discovered they'd be able to solve one of the world's most pressing problems—the shortage of energy. By putting a giant solar collector in orbit, sunlight could be converted into electricity and beamed down to Earth. Without using polluting fossil fuels, the world could have all the energy it needed.

O'Neill was so profoundly influenced by this discovery that he has since devoted much of his career to making space colonization a reality. He founded the Space Studies Institute in Princeton, New Jersey, to sponsor research projects directed toward space colonization and space power generation, and to educate the public about these possibilities.

His imagination and enthusiasm have stimulated scientists and engineers around the world to begin attacking these problems seriously. One of the earliest designs the students came up with was of two giant cylinders that would be tied together and put into orbit. The cylinders would spin around their long axes to generate artificial gravity so the

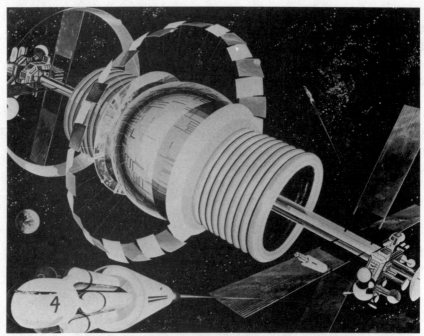

A NASA design for a space colony for 10,000 people. The people live in a sphere nearly a mile in circumference, which rotates to provide artificial gravity.

colonists could live inside them. They would have long windows to admit sunlight, and houses and crops could be situated within.

One of the delightful consequences of this design would be that if you climbed up to the center of the cylinders—the long axes around which they would spin—there would be no gravity. You could fly with wings, like Icarus in Greek mythology.

Another design that has been extensively studied looks like the space station in many an old science-fiction film. It would be wheel-shaped, a design often called the Stanford torus (doughnut), named for the research group that dreamed it up while assembled at Stanford University to study space colony designs. The wheel would rotate, making artificial gravity for the inhabitants within. The outside of the wheel would be covered with debris, perhaps from the surface of the moon, like sugar on a doughnut, shielding the inhabitants from solar flares.

The idea of living in space seems far-fetched to the average person, but in the 19th century, 19 million people immigrated to the U.S., suffering long, expensive, difficult journeys across thousands of miles of ocean to get here. They sought freedom, prosperity, and the chance to get away from the wars, prejudices, and decay of their own societies. The 21st century will, I expect, see similar numbers of people moving into space for the same old reasons.

There will be colonies of factories, colonies of Mormons and Seventh Day Adventists, vegetarians, Chinese, Socialists, Libertarians, and every other group looking for a new start. Earth is already overpopulated by many measures, and the population continues to increase, even if at a slower rate than before. Billions of people will be added in the next decades, and the yearning for a fresh start will sing in the hearts of many.

At first, it will be too expensive for most, but the costs will fall every year as new technologies and new ideas come along, just as the costs of air travel have fallen so much that, where only the rich could afford to travel by plane before World War II, now it's hard to find an American who hasn't flown at least once.

Several of the moons and planets can be transformed into livable places with the technology that should become available in future centuries. The first place to be converted— terraformed''—into a comfortable world may well be our own familiar moon.

At first glance, it looks as if the Moon is one world that could never have a home-like atmosphere. The Moon is too small to hold onto an atmosphere—that's why it's a vacuum today. But it was suggested many years ago that the Moon could be given an artificial atmosphere. The idea would be to put a power source on the Moon, or to beam sunlight to extract the oxygen from the minerals there and release it to begin the formation of an atmosphere.

It's true that the atmosphere will leak away eventually, but it might not do so for millions of years. The ingenious idea is that if we continue to supply the atmosphere, pumping it faster than it leaks away, then it doesn't matter if it eventually leaks into space. Over a period of many years, even centuries, a breathable atmosphere of oxygen could be built up on the Moon.

Undoubtedly, someone will file a gargantuan environmental impact statement exploring the effects the air will have, since that will then allow winds and weather to exist. However, taking into account the slow rotation of the Moon and the absence of large lakes or oceans

*TWELVE COMING TECHNOLOGICAL MILESTONES
IN SPACE*

Initial operation of a permanent Space Station;
*Inital operation of dramatically lower cost transport
vehicles to and from low Earth orbit for cargo and
passengers;*
*Addition of modular transfer vehicles capable of moving
cargoes and people from low Earth orbit to any destination
in the inner Solar System;*
A spaceport in low Earth orbit;
*Operation of an initial lunar outpost and pilot production
of rocket propellant;*
*Initial operation of a nuclear electric vehicle for high-
energy missions to the outer planets;*
First shipment of shielding mass from the Moon;
*Deployment of a spaceport in lunar orbit to support
expanding human operations on the Moon;*
*Initial operation of an Earth-Mars transportation system
for robotic precursor missions to Mars;*
*The first flight of a cycling spaceship to open continuing
passenger transport between Earth orbit and Mars orbit;*
*Human exploration and prospecting from astronaut out-
posts on Phobos, Deimos, and Mars; and*
*Start-up of the first Martian resource development base to
provide oxygen, water, food, construction materials, and
rocket propellants.*

—Pioneering the Space Frontier,
The Report of the National Commission on Space, 1986

of water, the weathering would be slow. So, before the reader runs
out to form an environmental conservation group to protect the Moon,
rest assured that the dangers to the lunar environment are small, and
the benefits to humanity are many.

An atmosphere would moderate the temperature extremes on the
Moon, producing an average temperature similar to Earth's. There
would be a greater range of temperatures because day and night are

each two weeks long. The Moon will heat up more during the lunar day than the Earth, and will cool down more at night. With an atmosphere, we could put water into small craters and form little lakes. (This possibility assumes that water is found on the Moon, or alternatively, imported from a comet or asteroid.)

Imagine swimming in the lunar lake. The gravity is one-sixth of Earth's, and you can breathe the new lunar atmosphere. You can jump six times higher than on the Earth. You can dive into the lake and watch the splash rise up, far overhead. Even if you are a terrible swimmer, you would be in no danger of drowning, because your body would be so light that it would hardly sink into the water. With simple floating shoes, you might even be able to walk on water.

In the weak lunar gravity, human-powered flight should be duck soup, almost as easy as in O'Neill's space colony, enabling any healthy person to fly with muscle-powered wings on the Moon.

Plants grow well in Moon dust, and it should not be long before we establish farms and parks there. The Moon will probably become the favorite recreation park of Earthlings, who will think no more of going to the Moon for a vacation than North Americans think of going to Europe.

Mars and Venus offer excellent prospects for terraforming into habitable worlds. Both already have atmospheres. Since they're carbon dioxide, they could be seeded with plants genetically engineered to fit local conditions, and the air could be converted into oxygen, in much the way it happened on Earth billions of years ago.

Initially, this would be done inside domes or tunnels, but over the centuries, we might decide to convert the entire planet's atmosphere into something breathable.

Mars already has water, in the form of ice, but Venus is a desert, so we'd have to import ice, perhaps from comets. One way or another, civilization will probably spread to those worlds.

What Now?

Momentum is building for an international human expedition to Mars. There is more support for such a joint mission in the U.S. and U.S.S.R. than ever before, and its major benefit to humanity might well be a reduction of tensions and a relief from the threat of world war.

The Russians know full well that a nation that masters the use of space will have the same global power that the great seafaring nations of old had. The Japanese, the Western Europeans, and the Chinese also know that a command of space travel is essential to future growth, and they are determined not to be left out.

The great seafaring nations—Britain, Spain, Portugal, France, Italy, Holland—owed much of their commercial power and cultural growth to the exploration and exploitation of resources around the world, permitted by their ability to travel vast distances in their vessels.

What are the three things we can do in the future? The first . . . is genetic engineering. What I predict . . . is that before long [we will say,] "Why did those people build spacecraft instead of growing them?" It would be much easier in the long run to use biological organization to build a spacecraft: To program the DNA and let the creature grow by itself rather than building it laboriously, one little piece at a time. So one might think of the spacecraft of the year 2000 or 2100 at least, . . . that you young people . . . are going to be flying, as something alive. . . .

The second big step in technology which we are going to see is artificial intelligence. That's to say, the programming of computers in such a way as to be compatible with the organization of a brain. That you could have a creature like the hummingbird with a brain that weighs a few grams, which can in fact do all kinds of wonderful things. But we don't have access to the brain of a hummingbird because we're not smart enough to program computers in such a way to be compatible. . . . Well, I think it is pretty clear that in twenty to thirty years we shall. So we will be able to build a machine, which is sort of a hybrid symbiosis of a plant, animal and electronic computer, the three components all working together, and information flowing back and forth between them.

The third new technology which we don't yet have . . . is just better propulsion systems. Particularly, solar electrical systems.

If you put those three systems together, I would say you could imagine . . . a spacecraft which is as capable as

Voyager, which weighs a kilogram instead of a couple of tons. That of course will make an enormous difference. We shall have then the possibility of getting to Uranus much quicker with a solar electrical system. . . .

What I think we should do is to think small and to think quick. That's the way space science should be going. We should be making our plans for the next 25 years on the assumption that we are not going to fly things like Voyager. Voyager was fine, but in a sense, that is the end of the technology of the 1900s. We're going to need radically different technology for the twenty-first century. . . .

The kind of spacecraft I envision is something that will get [to Uranus] in two years, and is slowed down by atmospheric braking, which for a light spacecraft is not too difficult. Then coast around, and with the residual sunlight, you would still have enough power to navigate from satellite to satellite. In the end, you will be carrying onboard something like a bombardier beetle. That's this famous creature which already developed, for its own nefarious purposes, a propulsion system which enables it to squirt scalding-hot gas at its enemies, so a little of that would enable you to hop around from satellite to satellite, or to go and graze on the rings of Uranus. Your imagination can take it on from there.

—Freeman Dyson, physicist, Institute for Advanced Study

Of course, much of their wealth was gained at the expense of the natives of other lands. Fortunately, there are no natives in the solar system whose lands may be stolen, so we have here a superb opportunity to bring wealth to our planet without hurting others. We can be sure that if the U.S. does not vigorously participate in this greatest of all adventures, plenty of others will.

Once we have obtained a foothold on other worlds, there will be no stopping human civilization. Even the destruction of a planet will not annihilate our species. We will continue to move on to other planets and moons, to asteroids and comets, and gradually we will bridge the gulf between the stars.

The end result will be prosperity and the answers to some of the greatest mysteries in the universe.

References

Angelo, Joseph A., Jr., *The Extraterrestrial Encyclopedia*, 1985, Facts on File, New York. Fairly technical book on all aspects of life in space.

Audouze, Jean, and Israel, Guy, *The Cambridge Atlas of Astronomy*, 1985, Cambridge University Press, New York.

Beatty, J. Kelly, O'Leary, Brian, and Chaikin, Andrew (eds.), *The New Solar System*, 2nd ed., 1982, Cambridge University Press, New York. Semi-technical.

Bell, Trudy E., *Upward: Status Report and Directory of the American Space Interest Movement, 1984-85*, available from the author at 11 Riverside Dr. #15GW, New York, NY 10023. The history and directory of pro-space groups.

Bova, Ben, *Star Peace: Assured Survival*, 1986, Tor, New York. Pro-"Star Wars."

Bowman, Robert M., *Star Wars*, 1986, Jeremy P. Tarcher, Los Angeles. Anti-"Star Wars."

Burrows, William E., *Deep Black*, 1986, Berkley Books, New York. Spy satellites.

Clark, David H., *Superstars*, 1984, McGraw-Hill, New York. Supernovas, pulsars, and such. Nontechnical.

Collins, Michael, *Liftoff*, 1988, Grove Press, New York. An astronaut's history of the space program, with thoughts on the future.

Cooper, Henry S.F., Jr., *A House in Space*, 1976, Holt Rinehart and Winston, New York. The story of our first Space Station, *Skylab*, and what it feels like to live in space.

Cooper, Henry S.F., Jr., *The Search for Life on Mars*, 1980, Holt Rinehart and Winston, New York. The story of the *Viking* landers.

Day, William, *Genesis on Planet Earth*, 1984, Yale University Press, New Haven, CT. The details of the origin of life. Technical.

Disney, Michael, *The Hidden Universe*, 1984, Macmillan, New York. The mystery of the universe's missing mass. Nontechnical.

Drexler, K. Eric, *Engines of Creation*, 1986, Anchor Press/Doubleday, New York. Pioneering, thought-provoking look at microtechnology and the revolutionary impact it will have on civilization.

Dyson, Freeman, *Infinite in All Directions*, 1988, Harper & Row, New York. A brilliant physicist's provocative thoughts about life, space, and the future.

Feinberg, Gerald, and Shapiro, Robert, *Life Beyond Earth*, 1980, William Morrow, New York. Nontechnical.

Finney, Ben R., and Jones, Eric M. (eds.), *Interstellar Migration and the Human Experience*, 1985, University of California Press, Berkeley, CA. An extraordinary assortment of views, from anthropologists to astronomers, on space colonization and human history.

Forward, Robert L., *Roundtrip Interstellar Travel Using Laser-Pushed Lightsails*, Vol. 21, No. 2, 1984, *Journal of Spacecraft*. Technical.

Forward, Robert L., *Starwisp: An Ultra-Light Interstellar Probe*, Vol. 21, No. 2, 1984, *Journal of Spacecraft*. Technical.

Forward, Robert L., and Davis, Joel, *Mirror Matter*, 1988, John Wiley & Sons, New York. The story of antimatter: what's going on and how it may be used in the future.

Friedman, Louis, *Starsailing*, 1988, John Wiley & Sons, New York. Solar sails in the solar system and beyond.

Goldsmith, Donald, *Nemesis*, 1985, Walker, New York. The theory that the dinosaurs were killed off by comets or meteorites. Nontechnical.

Goldsmith, Donald, *The Quest for Extraterrestrial Life*, 1980, University Science Books, Mill Valley, CA. A fascinating collection of most of the classic scientific papers on SETI. Most are technical, but many are quite readable.

Goldsmith, Donald, and Owen, Toby, *The Search for Life in the Universe*, 1980, Benjamin/Cummings, Menlo Park, CA. Nontechnical.

Greenstein, George, *Frozen Star*, 1983, New American Library, New York. Black holes and pulsars. Nontechnical.

Greenstein, George, *The Symbiotic Universe*, 1988, William Morrow, New York. How life arises in the universe.

Harwit, Martin, *Cosmic Discovery*, 1984, MIT Press, Cambridge, MA. Fascinating history of discovery and extrapolation to the future; technical.

Hawking, Stephen W., *A Brief History of Time*, 1988, Bantam Books, New York. A plain-English discussion of the bizarre nature of space and time, from brilliant physicist.

Heppenheimer, T.A., *Toward Distant Suns*, 1979, Fawcett Columbine, New York. One of the leading scientists in the field of space colonization tells how to do it.

Horowitz, Norman, *To Utopia and Back*, 1986, W.H. Freeman, San Francisco. The search for life in the solar system, by a *Viking* experimenter. Semi-technical.

Hyson, Michael T., "The Hy Tech Window," Suite 11, Box 305, 1200 S. Brand, Glendale, CA 91204. New newsletter from a brilliant, unconventional scientist, about space colonization, longevity, and other topics of the future.

Jastrow, Robert, *How to Make Nuclear Weapons Obsolete*, 1985, Little, Brown & Co., Boston. Pro-"Star Wars."

Joels, Mark Kerry, *The Mars One Crew Manual*, 1985, Ballantine Books, New York. Semi-factual handbook for an expedition to Mars.

Klass, Philip J., *The Public Deceived*, 1983, Prometheus Books, Buffalo, NY. Expose of UFO fakery.

Klass, Philip J., *UFO's Explained*, 1976, Random House, New York. Reasonable explanations for many important UFO cases.

Koval, Alexander, and Desinov, Lev, *Space Flights Serve Life on Earth*, 1987, Progress Publishers, Moscow. A Soviet point of view on the study of Earth from space.

Littman, Mark, *Planets Beyond*, 1988, John Wiley & Sons, New York. The outer planets: the history of their discovery and description of current research.

Long, F.A., Hafner, D., and Boutwell, J. (eds.), *Weapons in Space*, 1986, Norton, New York. Critical, technical look at "Star Wars."

Lovelock, James, and Allaby, Michael, *The Greening of Mars*, 1984, St. Martin's/Marek, New York. Colonizing Mars.

Lunan, Duncan, *Man and the Planets*, 1983, Ashgrove Press, Bath, England. Imaginative look at space exploration and colonization.

Mallove, Eugene T., *The Quickening Universe*, 1988, St. Martin's Press, New York. How life arises, and how it's being searched for.

Matloff, Gregory L., and Ubell, Charles, *World Ships: Prospects for Non-Nuclear Propulsion and Power Sources*, Vol. 38, pp. 253-261, 1985, *Journal of the British Interplanetary Society*. A design for a thousand-year interstellar colony-ship. Technical.

Matsunaga, Spark M., *The Mars Project*, 1986, Hill & Wang, New York. The senator looks at why we should go there.

McConnell, Malcolm, *Challenger: A Major Malfunction*, 1987, Double-day & Co., New York. What went wrong with the Space Shuttle.

McDonough, Thomas R., *The Search for Extraterrestrial Intelligence*, 1987, John Wiley & Sons, New York. Nontechnical.

Mendell, W. W., *Lunar Bases and Space Activities of the 21st Century*, 1985, Lunar and Planetary Institute, Houston. Numerous technical papers.

Mitton, Simon, *The Crab Nebula*, 1978, Charles Scribner's Sons, New York. Technical.

Moore, Patrick, and Hunt, Garry, *Atlas of the Solar System*, 1983, Rand McNally, Chicago. Beautifully illustrated.

Morrison, P., Billingham, J., and Wolfe, J., *The Search for Extraterrestrial Intelligence*, 1977, NASA SP-419; reprinted by Dover Publications, Mineola, NY. Technical.

Murray, Bruce, Malin, Michael C., and Greeley, Ronald, *Earthlike Planets*, 1981, W.H. Freeman, San Francisco. A comparison of Mercury, Venus, Earth, and Mars, with some discussion about exploring Mars in the future. Technical.

Muskie, Edmund S. (Ed.), *The U.S. In Space*, 1988, Center for National Policy Press, Washington, DC. Experts discuss the choices we face.

Myrabo, Leik, and Ing, Dean, *The Future of Flight*, 1985, Baen, New York. New types of propulsion becoming available, including lasers.

NASA, *Why Man Explores*, 1976, NASA Educational Publication 123, U.S. Government Printing Office, Washington, DC Scientists and writers ponder the question.

NASA, *Exploring the Living Universe*, 1988, US Government Printing Office, Washington, DC. The official NASA strategy for space life sciences.

NASA, *Spinoff 1988*, 1988, US Government Printing Office, Washington, DC. The annual series describing how life on Earth is enhanced by space technology.

National Academy of Sciences, *Space Science in the Twenty-First Century*, 1988, National Academy Press, Washington, DC. A distinguished body of scientists presents their plan.

National Commission on Space, *Pioneering the Space Frontier*, 1986, Bantam, New York. Fascinating, beautifully illustrated report on the President's commission, suggesting where we can and should go in the next 50 years.

Oberg, J.E., *Mission to Mars*, 1982, New American Library, New York. How we may go there.

Oberg, J.E., *New Earths*, 1981, Stackpole Books, Harrisburg, PA. How we can rebuild planets to make them nice places to live.

Oberg, J.E., and Oberg, Alcestis R., *Pioneering Space*, 1986, McGraw-Hill, New York. Our future on the final frontier.

Oberg, James E., *Unidentified Fraudulent Objects*, November, 1976, *Analog*. The UFOs that astronauts saw, identified.

Office of Technology Assessment, *Anti-Satellite Weapons, Countermeasures, and Arms Control*, 1985, U.S. Government Printing Office, Washington, DC.

Office of Technology Assessment, *Ballistic Missile Defense Technologies*, 1985, U.S. Government Printing Office, Washington, DC.

Office of Technology Assessment, *Civilian Space Stations and the U.S. Future in Space*, 1984, U.S. Government Printing Office, Washington, DC.

O'Leary, Brian, *The Making of an Ex-Astronaut*, 1970, Houghton Mifflin, Boston. Scientist-astronaut dropout.

O'Leary, Brian, *Project Space Station*, 1983, Stackpole Books, Harrisburg, PA. A nontechnical look at the Space Station by an ex-astronaut.

O'Neill, Gerard, *The High Frontier*, 1977, William Morrow, New York. The book that brought scientific respectability to the idea of space colonization.

Papagiannis, Michael D. (ed.), *The Search for Extraterrestrial Life: Recent Developments*, 1985, D. Reidel, Boston. Technical.

Pournelle, Jerry, and Ing, Dean, *Mutual Assured Survival*, 1984, Baen Enterprises, New York. Pro-"Star Wars."

Powers, Robert M., *Mars*, 1986, Houghton Mifflin, Boston. Nontechnical discussion of colonizing that planet.

Presidential Commission on the Space Shuttle Challenger Accident, *Report*, 5 vols., 1986, the White House, Washington, DC.

Ride, Sally K., *Leadership and America's Future in Space*, 1988, US Government Printing Office, Washington, DC. The "Ride Report," in which the astronaut outlines recommendations for NASA's goals from now until 2010.

Sagan, Carl, *Cosmos*, 1980, Random House, New York. An overview of astronomy, with consideration of SETI.

Sagan, Carl, *The Dragons of Eden*, 1977, Random House, New York. Speculations about the evolution of intelligence.

Sagan, Carl, Drake, F.D., Druyan, Ann, Ferris, Timothy, Lomberg, Jon, and Sagan, Linda Salzman, *Murmurs of Earth*, 1978, Ballantine Books, New York. The story of the *Voyager* message to the stars.

Schirra, Walter, and Billings, Richard N., *Schirra's Space*, 1988, Quinlan Press, Boston. The autobiography of one of our most experienced astronauts, with recommendations for the future.

Sheaffer, Robert, *The UFO Verdict*, 1981, Prometheus Books, Buffalo, NY. A critical study of UFOs.

Shapiro, Robert, *Origins*, 1986, Summit Books, New York. Nontechnical, critical examination of conflicting theories.

Shipman, Harry L., *Space 2000*, 1987, Plenum Press, New York. An alternative view of the near-term space program.

Shklovsky, I.S., and Sagan, Carl, *Intelligent Life in the Universe*, 1968, Dell Publishing, New York. The classic Russian-American book that helped bring respectability to SETI. Semi-technical.

Simon, Michael C., *Keeping the Dream Alive*, 1987, Earth Space Operations, San Diego. An aerospace executive discusses NASA, the commercialization of space, and future options, including colonization.

Stares, Paul B., *The Militarization of Space*, 1985, Cornell University Press, Ithaca, NY. A critical, scholarly study.

Stine, G. Harry, *The Third Industrial Revolution*, 1979, Ace, New York. How and why we will put industry in space.

Story, Ronald, *The Space-Gods Revealed*, 1976, Harper & Row, New York. Why von Däniken's *Chariots of the Gods* and the like are nonsense.

Trento, Joseph J., *Prescription for Disaster*, 1987, Crown Publishers, New York. A different analysis of what went wrong with the Space Shuttle.

Tsiolkovsky, K.E., *Selected Works*, 1968, Mir Publishers, Moscow. Some of his pioneering technical papers on astronautics, with a biography.

Tucker, Wallace and Karen, *The Dark Matter*, 1988, William Morrow & Co., New York. The story of one of the greatest mysteries in the universe: the case of the missing matter.

Washburn, Mark, *Mars at Last!* 1977, G.P. Putnam's Sons, New York. The story of the *Viking* landers.

Yeates, C.M., Johnson, T.V., Colin, L., Fanale, F.P., Frank, L., and Hunten, D.M., *Galileo*, 1985, NASA Special Publication 479, U.S. Government Printing Office, Washington, DC. Beautifully illustrated, semi-technical description of the *Galileo* spacecraft.

Organizations of Interest

Americans for the High Frontier, 2800 Shirlington Rd., Suite 405A, Arlington, VA 22206. Pro "Star Wars" and the civil space program.

Astronomical Society of the Pacific, 390 Ashton Ave., San Francisco, CA 94112. For amateurs and pros, even those far from the Pacific.

British Interplanetary Society, 27/29 South Lambeth Rd., London SW8 1SZ, England. For amateurs and pros.

Campaign for Space, Box 1526, Bainbridge, GA 31717. Political Action Committee.

Committee for the Scientific Investigation of Claims of the Paranormal, P.O. Box 229, Buffalo, NY 14215-0229. Critically investigates UFOs and such.

National Space Society, 922 Pennsylvania Ave. SE, Washington, DC 20003. Among other things, sponsors Dial-a-Shuttle, a number to call when a Space Shuttle flight is scheduled.

Ohio State University Radio Observatory, Ohio State University, Columbus, OH 43210. Needs money to continue the world's longest-running SETI system.

The Planetary Society, 65 N. Catalina Ave., Pasadena, CA 91106. Supports planetary exploration and SETI. The largest private space group in the solar system.

Royal Astronomical Society of Canada, 136 Dupont St., Toronto, Ontario M5R 1V2, Canada. Publishes a useful handbook of the sky.

SETI Institute, 101 First St., #410, Los Altos, CA 94022. Sponsors SETI research, including work on the NASA project.

Space Coalition, c/o Dickstein, Shapiro & Morin, 2101 L St., NW, Washington, DC 20037. Political Action Committee.

Spacepac, Suite 201-S, 3435 Ocean Park Blvd., Santa Monica, CA 90405. Political Action Committee.

Space Studies Institute, Box 82, Princeton, NJ 08542. Founded by space-colonization leader Gerard O'Neill, it supports colonization-oriented research projects.

Union of Concerned Scientists, 26 Church St., Cambridge, MA 02238. Anti "Star Wars," liberal group.

U.S. Space Education Assn., 746 Turnpike Rd., Elizabethtown, PA 17022. International nonprofit, nonpartisan group promoting peaceful space exploration.

World Space Foundation, Box Y, So. Pasadena, CA 91030. Supports asteroid search and solar-sail research.

Index